Selected Titles in This Series

(Continued in the back of this publication)

Wandering Vectors for Unitary Systems and Orthogonal Wavelets

MEMOIRS

of the
American Mathematical Society

Number 640

Wandering Vectors for Unitary Systems and Orthogonal Wavelets

Xingde Dai
David R. Larson

July 1998 • Volume 134 • Number 640 (end of volume) • ISSN 0065-9266

American Mathematical Society
Providence, Rhode Island

1991 *Mathematics Subject Classification.*
Primary 46N99, 47N40, 47N99;
Secondary 47D25, 47C05, 47D15, 46B28.

Library of Congress Cataloging-in-Publication Data

Dai, Xingde, 1946–
    Wandering vectors for unitary systems and orthogonal wavelets / Xingde Dai, David R. Larson.
      p. cm. — (Memoirs of the American Mathematical Society, ISSN 0065-9266 ; no. 640)
    "July 1998, volume 134, number 640 (end of volume)."
    Includes bibliographical references.
    ISBN 0-8218-0800-1 (alk. paper)
    1. Wavelets (Mathematics)   2. Operator theory.   3. Functional analysis.   I. Larson, David R.,
1942– . II. Title. III. Series.
QA3.A57 no. 640
510 s—dc21
[515′.2433]
                                     98-4219
                                        CIP

Memoirs of the American Mathematical Society

This journal is devoted entirely to research in pure and applied mathematics.

Subscription information. The 1998 subscription begins with volume 131 and consists of six mailings, each containing one or more numbers. Subscription prices for 1998 are $435 list, $348 institutional member. A late charge of 10% of the subscription price will be imposed on orders received from nonmembers after January 1 of the subscription year. Subscribers outside the United States and India must pay a postage surcharge of $30; subscribers in India must pay a postage surcharge of $43. Expedited delivery to destinations in North America $35; elsewhere $110. Each number may be ordered separately; *please specify number* when ordering an individual number. For prices and titles of recently released numbers, see the New Publications sections of the *Notices of the American Mathematical Society*.

Back number information. For back issues see the *AMS Catalog of Publications*.

Subscriptions and orders should be addressed to the American Mathematical Society, P. O. Box 5904, Boston, MA 02206-5904. *All orders must be accompanied by payment.* Other correspondence should be addressed to Box 6248, Providence, RI 02940-6248.

Copying and reprinting. Individual readers of this publication, and nonprofit libraries acting for them, are permitted to make fair use of the material, such as to copy a chapter for use in teaching or research. Permission is granted to quote brief passages from this publication in reviews, provided the customary acknowledgment of the source is given.

Republication, systematic copying, or multiple reproduction of any material in this publication (including abstracts) is permitted only under license from the American Mathematical Society. Requests for such permission should be addressed to the Assistant to the Publisher, American Mathematical Society, P. O. Box 6248, Providence, Rhode Island 02940-6248. Requests can also be made by e-mail to `reprint-permission@ams.org`.

Memoirs of the American Mathematical Society is published bimonthly (each volume consisting usually of more than one number) by the American Mathematical Society at 201 Charles Street, Providence, RI 02904-2294. Periodicals postage paid at Providence, RI. Postmaster: Send address changes to Memoirs, American Mathematical Society, P. O. Box 6248, Providence, RI 02940-6248.

Contents

ABSTRACT. We investigate topological and structural properties of the set $\mathcal{W}(\mathcal{U})$ of all complete wandering vectors for a system $\mathcal{U}$ of unitary operators acting on a Hilbert space. The special case of greatest interest is the system $\langle D, T \rangle$ of dilation (by 2) and translation (by 1) unitary operators acting on $L^2(\mathbb{R})$, for which the complete wandering vectors are precisely the orthogonal dyadic wavelets. The method we use is to parameterize $\mathcal{W}(\mathcal{U})$ in terms of a fixed vector ψ and the set of all unitary operators which locally commute with $\mathcal{U}$ at ψ. An analysis of the structure of this local commutant yields new information about $\mathcal{W}(\mathcal{U})$. The commutant of a unitary system can be abelian and yet the local commutant of it at a complete wandering vector can contain non-commutative von Neumann algebras as subsets. This is the case for $\langle D, T \rangle$. The unitary group of a certain non-commutative von Neumann algebra can be used to parameterize a connected class of wavelets generalizing those of Meyer with compactly supported Fourier transform.

The first author was supported in part by a grant from the AFOSR, by a YI grant from the Linear Analysis and Probability Workshop at Texas A&M University and by a grant provided by the University of North Carolina at Charlotte; The second author was supported in part by a grant from the NSF..

Received by the editor January 16, 1995.

Introduction

A *unitary system* is a set of unitary operators $\mathcal{U}$ acting on a Hilbert space $\mathcal{H}$ which contains the identity operator I of $\mathcal{B}(\mathcal{H})$. A *wandering vector* for $\mathcal{U}$ is a unit vector x with the property that

$$\mathcal{U}x := \{Ux : U \in \mathcal{U}\}$$

is an orthonormal set; it is called *complete* if $\mathcal{U}x$ is an orthonormal basis for $\mathcal{H}$. A *wandering vector system* will mean a unitary system which has a complete wandering vector.

Let $\mathcal{W}(\mathcal{U})$ denote the set of complete wandering vectors for a unitary system $\mathcal{U}$. For $\mathcal{W}(\mathcal{U})$ to be nonempty, the set $\mathcal{U}$ must be very special. It must be countable if it acts separably, and it must be discrete in the strong operator topology (pointwise convergence) because if $U, V \in \mathcal{U}$ and if x is a wandering vector for $\mathcal{U}$ then

$$\|U - V\| \geq \|Ux - Vx\| = \sqrt{2}.$$

Certain other properties are forced on $\mathcal{U}$ by the presence of a wandering vector. One purpose of this paper is to study such properties. Indeed, it was a matter of some surprise to us to discover that such a theory is viable even in some considerable generality. A more immediate purpose, however, is to study structural properties of $\mathcal{W}(\mathcal{U})$ for special systems $\mathcal{U}$ which are relevant to *wavelet* theory.

In operator theory, wandering vectors have been studied for groups of unitaries and semigroups of isometries, particularly those that are singly generated (c.f. [**13**]). Wavelet theory entails the study of wandering vectors for unitary systems which are not even semigroups.

In the past ten years wavelet theory has undergone a vast development, and many different aspects of the theory have been studied extensively in the literature. Some of the most frequently studied and used aspects include the following definition of orthogonal (or orthonormal) wavelet:

> *An orthogonal wavelet is a unit vector* $\psi(t)$ *in* $L^2(\mathbb{R}, \mu)$ *, with* μ
> *Lebesgue measure, such that* $\{2^{\frac{n}{2}} \psi(2^n t - l) : n, l \in \mathbb{Z}\}$ *constitutes*
> *an orthonormal basis for* $L^2(\mathbb{R})$*.*

This is the definition given in, e.g., Chui's book ([**3**], p.4), and it is referred to in Meyer's book ([**21**], p.28) as the Franklin-Stromberg definition. The simplest function satisfying this is the Haar wavelet $\psi_H = \chi_{[0,\frac{1}{2})} - \chi_{[\frac{1}{2},1)}$. Wavelets having this property (see, e.g., [**8**]) include those of Stromberg, Meyer, Battle-LeMarié and Daubechies. These wavelets also satisfy strong auxiliary regularity (differential and moment) properties and time-frequency localization properties which make them useful in applications. Dilation factors other than 2 (the dyadic case) have also been studied. Generalizations to $\mathbb{R}^n$ with matrix dilations, and $L^p(\mathbb{R}, \mu)$ for other

$1 \leq p < \infty$, are also frequently studied. Representative articles are contained in the excellent collections $[\mathbf{2},\ \mathbf{4}]$ and $[\mathbf{9}]$.

The term "mother wavelet" is also used in the literature for a function ψ satisfying the above definition of orthogonal wavelet. In this case the functions $\psi_{n,l} := 2^{\frac{n}{2}}\psi(2^n t - l)$ are called elements of the wavelet basis generated by the "mother." The functions $\psi_{n,l}$ will not themselves be mother wavelets unless $n = 0$.

The idea of viewing orthogonal wavelets as wandering vectors for dilation-translation unitary systems is simple and has been used by others (c.f. $[\mathbf{11},\ \mathbf{12}]$). Let T and D be the operators on $\mathcal{H} = L^2(\mathbb{R})$ defined by

$$(Tf)(t) = f(t - 1) \quad \text{and} \quad (Df)(t) = \sqrt{2}f(2t), f \in L^2(\mathbb{R}), t \in \mathbb{R}.$$

These are unitary operators, and are in fact bilateral shifts of infinite multiplicity, with wandering subspaces $L^2[0, 1]$ and $L^2([-2, -1] \cup [1, 2])$, respectively, considered as subspaces of $L^2(\mathbb{R})$. They fail to commute, but satisfy the relation $TD = DT^2$. For $\psi \in L^2(\mathbb{R})$ we have $2^{\frac{n}{2}}\psi(2^n t - l) = (D^n T^l \psi)(t)$.

The group generated by $\{D, T\}$ is easily computed to be

$$\text{Group}\{D, T\} = \{D^n T_\beta : n \in \mathbb{Z}, \beta \in \mathbb{D}\},$$

where $\mathbb{D}$ denotes the set of dyadic rational numbers, and where for real β, T_β denotes the translation unitary

$$(T_\beta f)(t) = f(t - \beta).$$

This can be viewed as a semidirect product of the dyadics by an action of the integers.

Every countable group $\mathcal{G}$ has a representation on a separable Hilbert space for which it has a wandering vector: just represent $\mathcal{G}$ on $l^2(\mathcal{G})$ by left multiplication, and note that $\chi_{\{g\}} \in \mathcal{W}(\mathcal{G})$ for $g \in \mathcal{G}$. For a different representation π, $\pi(\mathcal{G})$ may or may not have a wandering vector. In particular, this fails for Group$\{D, T\}$. To see this, choose $\beta_n \in \mathbb{D} \setminus \{0\}$, $\beta_n \to 0$, so then $T_\beta \to I$ in the strong operator topology (pointwise convergence). So if $\psi \in \mathcal{H}$ then $T_{\beta_n}\psi \to \psi$, but if ψ were wandering for Group$\{D, T\}$, then

$$\|T_{\beta_n}\psi - \psi\| = \sqrt{2} \text{ for all } n,$$

a contradiction. However, if we consider the subset

$$\mathcal{U}_{D,T} := \{D^n T^l : n, l \in \mathbb{Z}\},$$

then the property of ψ being a complete wandering vector for $\mathcal{U}_{D,T}$ is precisely that of ψ satisfying the definition of orthogonal wavelet. So $\mathcal{W}(\mathcal{U}_{D,T})$ is far from empty. It will sometimes be convenient to use the ordered pair notation $\langle D, T \rangle$ to abbreviate $\mathcal{U}_{D,T}$. Elements of $\mathcal{W}(\mathcal{U}_{D,T})$ will be called orthogonal wavelets.

It is useful for perspective to note that the *reversed* set $\mathcal{U}_{T,D} = \{T^n D^l : n, l \in \mathbb{Z}\}$ *fails* to have a wandering vector. To see this choose dyadic $\beta_k \to 0, \beta_k \neq 0$, as above. Write $\beta_{l_k} = p_k/2^{q_k}$, and note that (see Lemma 3.2) $D^{-q_k}T_{\beta_k} = T^{p_k}D^{-q_k}$. If $\psi \in \mathcal{H}$ then,

$$\|T^{p_k}D^{-q_k}\psi - D^{-q_k}\psi\| = \|D^{-q_k}T_{\beta_k}\psi - D^{-q_k}\psi\| = \|T_{\beta_k}\psi - \psi\| \to 0.$$

So if $\psi \in \mathcal{W}(\mathcal{U}_{T,D})$, then orthonormality of $\mathcal{U}_{T,D}\psi$ implies $T^{p_k} = I$ for all but finitely many k, contradicting the assumption that $\beta_k \neq 0$. (We thank Shijin Lu for this argument.)

Let $\mathcal{U}$ be a unitary system. The apparently new idea which we will develop in this article is an association of $\mathcal{W}(\mathcal{U})$ with the *commutant* of $\mathcal{U}$ in $\mathcal{B}(\mathcal{H})$, and more importantly, with the *local commutant* (see Chapter 1) of $\mathcal{U}$ at a complete wandering vector. This permits a partial analysis (and theoretically a complete analysis) of $\mathcal{W}(\mathcal{U})$ using operator-theoretic techniques. The most basic types of problems seem to be the topological ones: For a given unitary system $\mathcal{U}$, is the set $\mathcal{W}(\mathcal{U})$ closed in the Hilbert space norm topology? Is the linear span of $\mathcal{W}(\mathcal{U})$ dense in $\mathcal{H}$? Is $\mathcal{W}(\mathcal{U})$ norm-pathwise connected? Work on these issues lead to other types of operator-theoretic techniques and results. In the case of wavelets, these techniques do lead to the construction of new ones. The work we present is intended to be an initial step in a new direction.

We give a number of open problems within the context of our subject matter, and label by capital Roman letters A, $\cdots$, F those which we feel are the most significant.

Several graduate students who were enrolled in a topics course on this subject at Texas A&M University in the spring of 1994 contributed useful examples to this work. These are credited in context. Others made useful comments. We thank Eugen Ionascu, Vishnu Kamat, Shijin Lu, Darrin Speegle and Puhong You for their contributions.

Our basic analysis references for this article are [**10, 13, 14, 5, 16, 22**]; our basic wavelet references are [**3, 8, 20**].

We use $[\,\cdot\,]$ to denote closed linear span. If $\mathcal{S}$ is a set of operators, we use $\mathbb{U}(\mathcal{S})$ to denote the set of unitary operators in $\mathcal{S}$, and $\text{w}^*(\mathcal{S})$ for the von Neumann algebra generated by $\mathcal{S}$ and I. If $\mathcal{S}$ is a linear space of operators, a vector $x \in \mathcal{H}$ is called *cyclic* for $\mathcal{S}$ if $[\mathcal{S}x] = \mathcal{H}$, and x is *separating* for $\mathcal{S}$ if the map $A \to Ax$, $\mathcal{S} \to \mathcal{H}$, is injective. The notation $\mathcal{S}'$ will denote the *commutant* of $\mathcal{S}$: the set of operators in $\mathcal{B}(\mathcal{H})$ which commute with all elements in $\mathcal{S}$. In this article Hilbert spaces will be separable.

CHAPTER 1

The Local Commutant

Let $S \subseteq \mathcal{B}(\mathcal{H})$ be a set of operators, and let $x \in \mathcal{H}$ be a nonzero vector. Let us define
$$\mathcal{C}_x(S) := \{A \in \mathcal{B}(\mathcal{H}) : (AS - SA)x = 0, S \in S\}.$$
We call this the local (or "point") commutant of S at x. It can be a useful concept, especially when x is a cyclic vector for the linear span of S and S is not a semigroup. (If S is a semigroup and x is cyclic it reduces to the commutant by item (ii) below.) It is clearly a linear subspace of $\mathcal{B}(\mathcal{H})$ which is closed in the strong operator topology and the weak operator topology, and it contains the commutant S' of S. In the wavelet case, it turns out that the local commutant of $\mathcal{U}_{D,T}$ at a wavelet can contain non-abelian von Neumann algebras as subsets, and thus has rich structure, while the commutant itself is abelian. We capture some immediate and useful properties in the form of a lemma. Many of these generalize analogous properties of the commutant.

Lemma 1.1. *If $S \subseteq \mathcal{B}(\mathcal{H})$ is a set and if $x \in \mathcal{H}$ is a vector for which $[Sx] = \mathcal{H}$, then:*

1. *The vector x is separating for $\mathcal{C}_x(S)$.*
2. *If S is a semigroup then $\mathcal{C}_x(S) = S'$.*
3. *If A is an element of $\mathcal{C}_x(S)$ with dense range, then Ax is also cyclic for $[S]$.*
4. *Suppose x is also separating for S. Then if $S, T \in S$ with $ST \in S$ and $TS \in S$, and $ST \neq TS$, then neither S nor T is in $\mathcal{C}_x(S)$.*
5. *Suppose $S = S_1 S_2$ where S_1 is a semigroup. Then $\mathcal{C}_x(S) \subseteq S_1'$*
6. *If $V \in \mathcal{C}_x(S)$ is invertible, then*
$$\mathcal{C}_{Vx}(S) = \mathcal{C}_x(S)V^{-1}.$$

Proof. (i) If $A \in \mathcal{C}_x(S)$ and if $Ax = 0$, then for any $S \in S$ we have $ASx = SAx = 0$. So $ASx = 0$, hence $A = 0$.

(ii) The inclusion " $\supseteq$ " is trivial. For " $\subseteq$ ", suppose $A \in \mathcal{C}_x(S)$. Then for each $S, T \in S$ we have $ST \in S$, and so
$$AS(Tx) = (ST)Ax = S(ATx) = SA(Tx).$$
So since $T \in S$ was arbitrary and $[Sx] = \mathcal{H}$, it follows that $AS = SA$.

(iii) Then $SAx = ASx$, so the conclusion is immediate.

(iv) If $S \in \mathcal{C}_x(S)$, then since $T \in S$ we have $(ST - TS)x = 0$, so $STx = TSx$, which contradicts the fact that x separates S. Similarly, T cannot be in $\mathcal{C}_x(S)$.

(v) Then $\mathcal{S}_1\mathcal{S} \subseteq \mathcal{S}$. Let $A \in \mathcal{C}_x(\mathcal{S})$ and $R \in \mathcal{S}_1$. For each $S \in \mathcal{S}$ we have $ASx = SAx$, and also $ARSx = RSAx$ since $RS \in \mathcal{S}$. So $(AR - RA)Sx = 0$ for all $S \in \mathcal{S}$. Since $[\mathcal{S}x] = \mathcal{H}$, this implies that $AR = RA$, as required.

(vi) We have

$$
\begin{aligned}
\mathcal{C}_{Vx}(\mathcal{S}) &= \{A : (AS - SA)Vx = 0, S \in \mathcal{S}\} \\
&= \{A : (AVS - SAV)x = 0, S \in \mathcal{S}\} \\
&= \{A : AV \in \mathcal{C}_x(\mathcal{S})\} \\
&= \mathcal{C}_x(\mathcal{S}) \cdot V^{-1}. \; \square
\end{aligned}
$$

If $\mathcal{S}$ *contains* a semigroup $\mathcal{S}_0$ with $\mathcal{S}_0' = \mathcal{S}$, and if x is cyclic for $\mathcal{S}_0$ in the sense that $[\mathcal{S}_0 x] = \mathcal{H}$, then part (ii) of the above lemma implies that $\mathcal{C}_x(\mathcal{S})$ reduces to simply $\mathcal{S}'$. This is the case for $\mathcal{U}_{D,T}$ when x is a *scaling* function. (See §6.3.) However, if x is an orthogonal *wavelet* the structure of $\mathcal{C}_x(\mathcal{U}_{D,T})$ is much richer, as we will see.

The following is included only for perspective. It shows that local commutants and commutants share an additional special structural property.

Proposition 1.2. *Let* $\mathcal{S} \subseteq \mathcal{B}(\mathcal{H})$ *and* $x \in \mathcal{H}$ *be arbitrary. Then* $\mathcal{C}_x(\mathcal{S})$ *is* 2-*reflexive in the sense that* $\mathcal{C}_x(\mathcal{S}) \otimes I_2$ *is reflexive in* $\mathcal{B}(\mathcal{H} \otimes \mathcal{H}_2)$.

Proof. We use a duality proof. From [17] it will be enough to show that the preannihilator of $\mathcal{C}_x(\mathcal{S})$ in $\mathcal{C}_1(\mathcal{H})$ is generated by operators of rank ≤ 2. We claim in fact that

$$
(\mathcal{C}_x(\mathcal{S}))_\perp = \overline{\mathrm{span}}\{[S, x \otimes y] : S \in \mathcal{S}, y \in \mathcal{H}\},
$$

where for $x, y \in \mathcal{H}, x \otimes y$ denotes the rank-1 operator defined by

$$
(x \otimes y)w := \langle w, y \rangle x, w \in \mathcal{H},
$$

and

$$
[S, x \otimes y] := S(x \otimes y) - (x \otimes y)S.
$$

For arbitrary $A \in \mathcal{B}(\mathcal{H})$ we have the tracial equation

$$
\begin{aligned}
\mathrm{Tr}(A[S, x \otimes y]) &= \mathrm{Tr}(A(Sx \otimes y - x \otimes S^*y)) \\
&= \mathrm{Tr}(ASx \otimes y) - \mathrm{Tr}(Ax \otimes S^*y) \\
&= \langle ASx, y \rangle - \langle Ax, S^*y \rangle \\
&= \langle ASx, y \rangle - \langle SAx, y \rangle \\
&= \langle (AS - SA)x, y \rangle.
\end{aligned}
$$

From this it follows that $A \in \mathcal{C}_x(\mathcal{S})$ iff A is annihilated by all trace class operators of the form $[S, x \otimes y]$ for $S \in \mathcal{S}$ and $y \in \mathcal{H}$. So $(\mathcal{C}_x(\mathcal{S}))_\perp$ is the $\|\cdot\|_1$-closed linear span of these commutators. Since each of these has rank ≤ 2, this proves that $\mathcal{C}_x(\mathcal{S})$ is 2-reflexive. $\square$

The key to our approach is the following simple result.

Proposition 1.3. *Let* $\mathcal{U}$ *be a unitary system in* $\mathcal{B}(\mathcal{H})$. *Suppose* $\psi \in \mathcal{W}(\mathcal{U})$. *Then*

$$
\mathcal{W}(\mathcal{U}) = \{V\psi : V \in \mathbb{U}(\mathcal{C}_\psi(\mathcal{U}))\}.
$$

Moreover, the correspondence

$$V \to V\psi, \ \mathbb{U}(\mathcal{C}_\psi(\mathcal{U})) \to \mathcal{W}(\mathcal{U}),$$

is one-to-one.

Proof. Let $V \in \mathbb{U}(\mathcal{C}_\psi(\mathcal{U}))$. Let $\eta = V\psi$. For $U \in \mathcal{U}$ we have

$$U\eta = UV\psi = VU\psi$$

since U commutes with V at ψ. Thus $\mathcal{U}\eta = V\mathcal{U}\psi$, and so $\mathcal{U}\eta$ is an orthonormal basis for $\mathcal{H}$ since V is unitary. So $\eta \in \mathcal{W}(\mathcal{U})$.

Conversely, let $\eta \in \mathcal{W}(\mathcal{U})$ be arbitrary. Since $\mathcal{U}\psi$ and $\mathcal{U}\eta$ are orthonormal bases, there is a unique unitary operator V with

$$VU\psi = U\eta, \ U \in \mathcal{U}.$$

Then $V\psi = \eta$ since $I \in \mathcal{U}$. So $VU\psi = UV\psi$ for all $U \in \mathcal{U}$. Thus $V \in \mathcal{C}_\psi(\mathcal{U})$.

By Lemma 1.1, ψ separates points of $\mathcal{C}_\psi(\mathcal{U})$. Thus the map $V \to V\psi$ is one-to-one. $\square$

Proposition 1.3 shows that if $\mathcal{U}$ is a unitary system with $\mathcal{W}(\mathcal{U}) \neq \emptyset$, then given any $\psi \in \mathcal{W}(\mathcal{U})$ the entire set $\mathcal{W}(\mathcal{U})$ can be parameterized in a natural way by the set of unitary operators in the local commutant of $\mathcal{U}$ at ψ.

It is an elementary result that if $\mathcal{G} \subset \mathcal{B}(\mathcal{H})$ is a unitary group, and if x and y are cyclic vectors for span $\mathcal{G}$, then the vector functionals

$$\omega_x = \langle \cdot x, x \rangle \ \text{and} \ \omega_y = \langle \cdot y, y \rangle$$

agree on $\mathcal{G}$ iff there exists $V \in \mathbb{U}(\mathcal{G}')$ with $Vx = y$. Proposition 1.3 can be thought of as a special case of the following generalization of this to unitary systems.

Proposition 1.4. *Let $\mathcal{U}$ be a unitary system in $\mathcal{B}(\mathcal{H})$. Suppose $x, y \in \mathcal{H}$ with $[\mathcal{U}x] = [\mathcal{U}y] = \mathcal{H}$. Then*

$$\langle U_1 x, U_2 x \rangle = \langle U_1 y, U_2 y \rangle$$

for all $U_1, U_2 \in \mathcal{U}$ if and only if there is a unitary $V \in \mathcal{C}_x(\mathcal{U})$ with $Vx = y$.

Proof. Fix $U_1, U_2, \cdots, U_n \in \mathcal{U}$ and $\lambda_1, \lambda_2, \cdots \lambda_n \in \mathbb{C}$. Then

$$\left\langle \sum \lambda_i U_i x, \sum \lambda_i U_i x \right\rangle = \sum_{i,j} \overline{\lambda}_j \lambda_i \langle U_j^* U_i x, x \rangle$$

$$= \sum_{i,j} \overline{\lambda}_j \lambda_i \langle U_j^* U_i y, y \rangle$$

$$= \left\langle \sum \lambda_i U_i y, \sum \lambda_i U_i y \right\rangle.$$

This shows that the map

$$V : \text{span}(\mathcal{U}x) \to \text{span}(\mathcal{U}y)$$

defined by $V(\sum \lambda_i U_i x) = \sum \lambda_i U_i y$ is isometric. Thus since $[\mathcal{U}x] = [\mathcal{U}y] = \mathcal{H}$, V extends to a unitary operator. For $U \in \mathcal{U}$ we have $VUx = Uy = UVx$. Thus $V \in \mathcal{C}_x(\mathcal{U})$. $\square$

Proposition 1.4 induces an equivalence relation on the set of cyclic vectors for span$(\mathcal{U})$ where in $\mathcal{W}(\mathcal{U})$ constitutes one equivalence class. There is, in addition, a natural equivalence relation on the set $\mathcal{W}(\mathcal{U})$ itself induced by the usual notion of the equivalence of group representations (see §6.1).

In certain cases new wandering vectors can be obtained by "interpolating" between a known pair. The following proposition can be viewed as the prototype of our results of chapter 5.

Proposition 1.5. *Let* $\mathcal{U}$ *be a unitary system, let* $\psi, \eta \in \mathcal{W}(\mathcal{U})$, *and let* V *be the unique unitary operator in* $\mathcal{C}_\psi(\mathcal{U})$ *with* $V\psi = \eta$. *Suppose* $V^2 = I$. *Then*

$$\cos \alpha \cdot \psi + i \sin \alpha \cdot \eta$$

is in $\mathcal{W}(\mathcal{U})$ *for all* $0 \le \alpha \le 2\pi$.

Proof. Let $P = \frac{1}{2}(V + I)$. Then P is a projection, and is contained in $\mathcal{C}_\psi(\mathcal{U})$. Let

$$\omega_1 = \cos \alpha + i \sin \alpha \quad \text{and} \quad \omega_2 = \cos \alpha - i \sin \alpha.$$

Then $|\omega_i| = 1$, so $W := \omega_1 P + \omega_2(I - P)$ is a unitary operator in $\mathcal{C}_\psi(\mathcal{U})$. So $W\psi \in \mathcal{W}(\mathcal{U})$. We have $W\psi = \omega_1 P\psi + \omega_2(I - P)\psi$, and $P = \frac{1}{2}(V + I)$, so

$$P\psi = \frac{1}{2}V\psi + \frac{1}{2}\psi = \frac{1}{2}(\eta + \psi).$$

Similarly

$$(I - P)\psi = \frac{1}{2}(\psi - \eta).$$

Thus $W\psi = \frac{1}{2}(\omega_1 + \omega_2)\psi + \frac{1}{2}(\omega_1 - \omega_2)\eta = \cos \alpha \cdot \psi + i \sin \alpha \cdot \eta$. $\square$

Unitaries V with $V^2 = I$ are called involutions, or symmetries. It turns out that the involutive case in Proposition 1.5 is not uncommon in dilation-translation wavelet theory (Chapter 5.) Such pairs (ψ, η) are of course connected in $\mathcal{W}(\mathcal{U})$ in the Hilbert space metric.

If $\mathcal{U}$ is a unitary system which is not a group, and if $\mathcal{W}(\mathcal{U}) \ne \emptyset$, then $\mathcal{U}$ is not even a semigroup.

Lemma 1.6. *Let* $\mathcal{S}$ *be a unital semigroup of unitaries in* $\mathcal{B}(\mathcal{H})$. *Suppose* $\mathcal{W}(\mathcal{S}) \ne \emptyset$. *Then* $\mathcal{S}$ *is a group.*

Proof. Let $\psi \in \mathcal{W}(\mathcal{S})$. If $\mathcal{S}$ is not a group, let $U \in \mathcal{S}$ such that $U^{-1} \notin \mathcal{S}$. Then for each $V \in \mathcal{S}$,

$$\langle U^{-1}\psi, V\psi \rangle = \langle \psi, UV\psi \rangle = 0,$$

since $UV \in \mathcal{S}$ and $V \ne U^{-1}$. Hence $U^{-1}\psi$ is a non-zero vector orthogonal to $[\mathcal{S}\psi]$, a contradiction. $\square$

If $\mathcal{U}$ is a unitary system then $\mathcal{C}_\psi(\mathcal{U})$ is almost never an algebra, and its set of unitary operators is almost never a group.

Proposition 1.7. *Let* $\mathcal{U}$ *be a unitary system, and suppose* $[\mathcal{W}(\mathcal{U})] = \mathcal{H}$. *Then*

1. *If* $\mathcal{C}_\psi(\mathcal{U})$ *is an algebra for some* $\psi \in \mathcal{W}(\mathcal{U})$, *then* $\mathcal{C}_\eta(\mathcal{U}) = \mathcal{U}'$ *for every* $\eta \in \mathcal{W}(\mathcal{U})$. *In particular,* $\mathcal{C}_\eta(\mathcal{U})$ *is an algebra for all* $\eta \in \mathcal{W}(\mathcal{U})$.
2. *If* $\mathbb{U}(\mathcal{C}_\psi(\mathcal{U}))$ *is a semigroup for some* $\psi \in \mathcal{W}(\mathcal{U})$, *then* $\mathbb{U}(\mathcal{C}_\eta(\mathcal{U})) = \mathbb{U}(\mathcal{U}')$ *for every* $\eta \in \mathcal{W}(\mathcal{U})$. *In particular,* $\mathbb{U}(\mathcal{C}_\psi(\mathcal{U}))$ *is a group for all* $\eta \in \mathcal{W}(\mathcal{U})$.

Proof. Let $\eta \in \mathcal{W}(\mathcal{U})$ be arbitrary. By Proposition 1.3, there is a unique $V \in \mathbb{U}(\mathcal{C}_\eta(\mathcal{U}))$ with $\psi = V\eta$. Then

$$\mathcal{C}_\psi(\mathcal{U}) = \mathcal{C}_\eta(\mathcal{U})V^*$$

by Lemma 1.1 (vi). Hence $V^* \in \mathcal{C}_\psi(\mathcal{U})$.

For item (i), if $\mathcal{C}_\psi(\mathcal{U})$ is closed under multiplication then $\mathcal{C}_\psi(\mathcal{U})V^* \subseteq \mathcal{C}_\psi(\mathcal{U})$, and it follows that

$$\mathcal{C}_\psi(\mathcal{U}) \subseteq \mathcal{C}_\eta(\mathcal{U}).$$

So if $A \in \mathcal{C}_\psi(\mathcal{U})$ then $(AU - UA)\eta = 0$ for all $U \in \mathcal{U}$ and for all $\eta \in \mathcal{W}(\mathcal{U})$. Since $[\mathcal{W}(\mathcal{U})] = \mathcal{H}$, this implies $A \in \mathcal{U}'$. We have shown $\mathcal{C}_\psi(\mathcal{U}) = \mathcal{U}'$. Again let $\eta \in \mathcal{W}(\mathcal{U})$ be arbitrary, and let W be the unique unitary in $\mathcal{C}_\psi(\mathcal{U}) = \mathcal{U}'$ with $\eta = W\psi$. Then

$$\mathcal{C}_\eta(\mathcal{U}) = \mathcal{C}_\psi(\mathcal{U})W^* = \mathcal{U}'.$$

For item (ii), if $\mathbb{U}(\mathcal{C}_\psi(\mathcal{U}))$ is a semigroup then $\mathbb{U}(\mathcal{C}_\psi(\mathcal{U}))V^* \subseteq \mathbb{U}(\mathcal{C}_\psi(\mathcal{U}))$. Since also $\mathbb{U}(\mathcal{C}_\psi(\mathcal{U})) = \mathbb{U}(\mathcal{C}_\eta(\mathcal{U}))V^*$ this implies

$$\mathbb{U}(\mathcal{C}_\psi(\mathcal{U})) \subseteq \mathbb{U}(\mathcal{C}_\eta(\mathcal{U})).$$

If $S \in \mathbb{U}(\mathcal{C}_\psi(\mathcal{U}))$ then

$$(SU - US)\eta = 0, \; U \in \mathcal{U}, \; \eta \in \mathcal{W}(\mathcal{U}),$$

so as above, $S \in \mathcal{U}'$. The rest is identical to (i). $\square$

Proposition 1.8 . *Let $\mathcal{U}$ be a unitary system, and suppose $\mathcal{C}_\psi(\mathcal{U})$ is abelian for some $\psi \in \mathcal{W}(\mathcal{U})$. Then $\mathcal{C}_\eta(\mathcal{U})$ is abelian for all $\eta \in \mathcal{W}(\mathcal{U})$.*

Proof. Suppose $\mathcal{C}_\psi(\mathcal{U})$ is abelian and let $\eta \in \mathcal{W}(\mathcal{U})$ be arbitrary. Let $V \in \mathbb{U}(\mathcal{C}_\eta(\mathcal{U}))$ with $\psi = V\eta$. Then $\mathcal{C}_\psi(\mathcal{U}) = \mathcal{C}_\eta(\mathcal{U})V^*$. So $V^* \in \mathcal{C}_\psi(\mathcal{U})$. Since $V^* \in (\mathcal{C}_\psi(\mathcal{U}))'$ and V is normal, $V \in (\mathcal{C}_\psi(\mathcal{U}))'$. So $\mathcal{C}_\eta(\mathcal{U}) = \mathcal{C}_\psi(\mathcal{U})V$ is abelian. $\square$

Example 1.9. Let $\{e_n\}_{-\infty}^\infty$ be an orthonormal basis for a separable Hilbert space $\mathcal{H}$, and let $Se_n = e_{n+1}$ be the bilateral shift of multiplicity one. Let $\mathcal{U} = \{S^n : n \in \mathbb{Z}\}$ be the group generated by S. Each e_n is in $\mathcal{W}(\mathcal{U})$. By Lemma 1.1 part (ii) and Proposition 1.3,

$$\mathcal{W}(\mathcal{U}) = \{Ve_0 : V \in \mathbb{U}(\{S\}')\}.$$

Here $\{S\}'$ coincides with the set of Laurent operators. Let $\mathbb{T}$ be the unit circle. If we represent S on $L^2(\mathbb{T})$ in the usual way by identifying it with the multiplication operator M_z, then $\mathbb{U}(\{S\}')$ is identified with (multiplication by) the set of unimodular functions on $\mathbb{T}$, and e_0 is identified with the constant function 1. Then Proposition 1.3 just recovers the well-known fact that the set of complete wandering vectors for the shift coincides (under this representation)with the set of unimodular functions on $\mathbb{T}$. In this case $\mathcal{W}(\mathcal{U})$ is clearly a closed, connected subset of the unit ball of $\mathcal{H}$ in the norm topology with dense linear span.

Example 1.10. Let $\mathcal{G}$ be a countable group, let $\mathcal{H} = l^2(\mathcal{G})$, and let π_L be the left regular representation of $\mathcal{G}$ on on $\mathcal{H}$. That is, for $h \in \mathcal{G}$ and $\{\lambda_g\}_{g \in \mathcal{G}} \in l^2(\mathcal{G})$, define $\pi_L(h)\{\lambda_g\} = \{\lambda_{h^{-1}g}\}$; so writing $\lambda(g) \equiv \{\lambda_g\}_{g \in \mathcal{G}}$, we have

$$
\begin{aligned}
(\pi_L(h_1)\pi_L(h_2)\lambda)(g) &= (\pi_L(h_1)\lambda(h_2^{-1}\cdot))(g) \\
&= (\lambda(h_2^{-1}\cdot))(h_1^{-1}g) \\
&= \lambda(h_2^{-1}h_1^{-1}g) \\
&= \lambda((h_1h_2)^{-1}g) \\
&= (\pi_L(h_1h_2)\lambda)(g)
\end{aligned}
$$

The standard basis for $\mathcal{H}$ is $\{e_g : g \in \mathcal{G}\}$, where $e_g = \chi_{\{g\}} \equiv \{\delta_{g,k}\}_{k \in \mathcal{G}}$. Then

$$\pi_L(h)e_g = \{\delta_{g,h^{-1}k}\}_{k \in \mathcal{G}} = \{\delta_{hg,k}\}_{k \in \mathcal{G}} = e_{hg}.$$

The vectors e_g are clearly in $\mathcal{W}(\pi_L(\mathcal{G}))$. By Lemma 1.1 the local commutant of $\pi_L(\mathcal{G})$ at e_I is just the commutant, (where I denotes the identity element of $\mathcal{G}$). Since $\{\pi_L(\mathcal{G})\}'$ is a von Neumann algebra, its group of unitaries is connected in the norm topology. Since the map $V \to Ve_I$ is continuous, Proposition 1.3 implies that $\mathcal{W}(\pi_L(\mathcal{G}))$ is a connected subset of the unit ball of $\mathcal{H}$. Since $e_g \in \mathcal{W}(\pi_L(\mathcal{G})), g \in \mathcal{G}$, the set $\mathcal{W}(\pi_L(\mathcal{G}))$ has dense span.

The algebra $\mathrm{w}^*(\pi_L(\mathcal{G}))$ considered above and its commutant are classic in the theory of von Neumann algebras. See, for instance, §6.7 in [14]. From this theory we have

$$\{\pi_L(\mathcal{G})\}' = \mathrm{w}^*(\pi_R(\mathcal{G})),$$

where $\pi_R(\mathcal{G})$ is the right regular representation of $\mathcal{G}$ on $\mathcal{H} = l^2(\mathcal{G})$ defined by $\pi_R(h)\{\lambda_g\} = \{\lambda_{gh}\}$. Moreover, if $\mathcal{G}$ is a group which has the property that the conjugacy class of each element other than the identity is infinite (an *i.c.c* group) then $\mathrm{w}^*(\pi_L(\mathcal{G}))$ and $\mathrm{w}^*(\pi_R(\mathcal{G}))$ are factor von Neumann algebras of type II_1. This is the case, for instance, if $\mathcal{G}$ is the free (non-abelian) group on n generators for $n \geq 2$. This shows that for some unitary systems $\mathcal{U}$ (and perhaps for many) the structure of $\mathcal{W}(\mathcal{U})$ is at least as complex as the structure of the unitary group of a type II_1 factor.

Example 1.11. Let $\mathcal{G}$ be a countable group, and let $\mathcal{G}_0 \subseteq \mathcal{G}$ be a *subset* containing the identity element I. Sometimes it is possible to obtain a faithful unitary representation π of $\mathcal{G}$ on $l^2(\mathcal{G}_0)$ satisfying the requirement that if $h \in \mathcal{G}$ and $g \in \mathcal{G}_0$ are such that $h^{-1}g \in \mathcal{G}_0$, then $\pi(h)e_g = e_{hg}$. (Where as in the previous example $e_g \equiv \chi_{\{g\}}$.) Then $e_I \in \mathcal{W}(\pi(\mathcal{G}_0))$ trivially.

This "example" is generic. If $\mathcal{U}$ is any unitary system on a Hilbert space $\mathcal{H}$ with a complete wandering vector ψ, let $\widetilde{\mathcal{U}}$ be the group generated by $\mathcal{U}$ in $\mathcal{B}(\mathcal{H})$. Let $\mathcal{G} = \widetilde{\mathcal{U}}$ as an abstract group, and let $\mathcal{G}_0 = \mathcal{U}$, a subset of $\mathcal{G}$. Let $\mathcal{K} = l^2(\mathcal{G}_0)$. Define a unitary operator

$$W : \mathcal{H} \to \mathcal{K} \quad \text{by} \quad Wg\psi = e_g, g \in \mathcal{U},$$

making use of the fact that $\mathcal{U}\psi$ is an orthonormal basis for $\mathcal{H}$. Define

$$\pi : \mathcal{B}(\mathcal{H}) \to \mathcal{B}(\mathcal{K}) \quad \text{by} \quad \pi(A) = WAW^*,$$

and restrict to $\mathcal{G} \equiv \widetilde{\mathcal{U}}$. Then π satisfies the property of the above paragraph, and is unitarily equivalent to the identity representation of $\mathcal{G} \equiv \widetilde{\mathcal{U}}$ on $\mathcal{H}$.

Let us call a unitary representation π of a group $\mathcal{G}$ relative to a unital subset $\mathcal{G}_0$ a *wandering vector representation* of the pair $(\mathcal{G}, \mathcal{G}_0)$ if it is faithful on $\mathcal{G}$ and if

$\pi(\mathcal{G}_0)$ has a complete wandering vector. From above, these are unitarily equivalent to those of Example 1.11. In the context of wandering vector theory, and especially wavelet theory, an abstract question which becomes rather intriguing is: Given a group $\mathcal{G}$, what are the unital subsets $\mathcal{G}_0$ which are *allowable* in the sense that $(\mathcal{G}, \mathcal{G}_0)$ has a wandering vector representation? In particular, if $\mathcal{G}$ is generated as a group by an ordered pair of elements $\{g_1, g_2\}$, and if $\mathcal{G}_i = \mathrm{Group}\{g_i\}$, is the set

$$\{\mathcal{G}_1 \mathcal{G}_2 = h_1 h_2 : h_i \in \mathcal{G}_i\}$$

an allowable subset of $\mathcal{G}$? Work here may aid in understanding wavelet systems, in particular. The question is obviously nontrivial in view of wavelet theory. This question generalizes to ordered n-tuples of generators. As mentioned earlier, we have the problems: when does $\mathcal{W}(\pi(\mathcal{G}_0))$ have dense span, when is it closed, and when is it connected?

To gain some insight we can abstract the $\langle D, T \rangle$ wavelet system in the introduction.

Example 1.12. (The abstract one-dimensional system.) Let $Se_n = e_{n+1}$, be the bilateral shift of multiplicity one in Example 1.9. Let

$$A = S \otimes I \in \mathcal{B}(\mathcal{H} \otimes \mathcal{H}),$$

and let B be any unitary in $\mathcal{B}(\mathcal{H} \otimes \mathcal{H})$ with

$$B|_{e_0 \otimes \mathcal{H}} = (I \otimes S)|_{e_0 \otimes \mathcal{H}}.$$

Let $\mathcal{U}_{A,B} := \{A^n B^l : n, l \in \mathbb{Z}\}$. Then $\mathcal{U} = \mathcal{U}_A \mathcal{U}_B$ where $\mathcal{U}_A$ and $\mathcal{U}_B$ are the groups generated by A and B, respectively. We have $e_0 \otimes e_n \in \mathcal{W}(\mathcal{U})$ for each $n \in \mathbb{Z}$, but except in special cases $e_p \otimes e_n$ will not be in $\mathcal{W}(\mathcal{U})$ if $p \neq 0$. Note that by Lemma 1.1 (v), if $\psi \in \mathcal{W}(\mathcal{U})$ then every operator in $\mathcal{C}_\psi(\mathcal{U})$ commutes with A. Also, by Lemma 1.1 (iv), if A and B do not commute, then neither A nor B can ever be in $\mathcal{C}_\psi(\mathcal{U})$.

The above example is generic in that it is really a "model" for "one-dimensional" wandering vector systems such as $\langle D, T \rangle$. To see this, let $\mathcal{K}$ be an *arbitrary* separable Hilbert space, and let U and V be *arbitrary* unitary operators in $\mathcal{B}(\mathcal{K})$. Let

$$\mathcal{U}_{U,V} = \{U^n V^l : n, l \in \mathbb{Z}\},$$

and suppose $\mathcal{W}(\mathcal{U}_{U,V})$ is nonempty. Let $\psi \in \mathcal{W}(\mathcal{U}_{U,V})$. Since the sets

$$\{U^n V^l \psi : n, l \in \mathbb{Z}\} \quad \text{and} \quad \{e_n \otimes e_l : n, l \in \mathbb{Z}\}$$

are orthonormal bases for $\mathcal{K}$ and $\mathcal{H} \otimes \mathcal{H}$, respectively, they determine a unitary operator $W : \mathcal{K} \to \mathcal{H} \otimes \mathcal{H}$ such that

$$WU^n V^l \psi = e_n \otimes e_l$$

for all n, l. Then $W\psi = e_0 \otimes e_0$, and for each $n, l \in \mathbb{Z}$ we have

$$
\begin{aligned}
WUW^*(e_n \otimes e_l) &= WUU^n V^l \psi \\
&= WU^{n+1} V^l \psi \\
&= e_{n+1} \otimes e_l \\
&= (S \otimes I)(e_n \otimes e_l).
\end{aligned}
$$

So
$$WUW^* = S \otimes I = A.$$

Let $B = WVW^*$. Then
$$
\begin{aligned}
B(e_0 \otimes e_l) &= WVW^*(e_0 \otimes e_l) \\
&= WVU^0 V^l \psi \\
&= WV^{l+1} \psi \\
&= e_0 \otimes e_{l+1} \\
&= (I \otimes S)(e_0 \otimes e_l)
\end{aligned}
$$

for each $l \in \mathbb{Z}$, so
$$B|_{e_0 \otimes \mathcal{H}} = (I \otimes S)|_{e_0 \otimes \mathcal{H}},$$

as required.

As above we have the closure, span and connectedness questions for $\mathcal{W}(\mathcal{U}_{A,B})$. For the wavelet system $\langle D, T \rangle$, Example 4.5(ii) shows that $\mathcal{W}(\mathcal{U}_{D,T})$ is *not* closed and Corollary 3.17 shows that span $\mathcal{W}(\mathcal{U}_{D,T})$ *is* dense. Are these properties true of general systems of the form $\mathcal{U}_{A,B}$? When is $\mathcal{W}(\mathcal{U}_{A,B})$ connected? We have no counterexample. On the other hand, examples are very hard to evaluate.

Problem A. *Is $\mathcal{W}(\mathcal{U}_{D,T})$ connected?*

This connectedness problem is perhaps the most important open question in our theory. A solution may lead to perturbation methods for wavelets, in particular. Construction of a counterexample for some other system $\mathcal{U}_{A,B}$ *may* shed some light on this matter. Based on evidence so far, we conjecture that the answer is "yes." There are related optimization problems, such as computation of $\text{dist}(x, \mathcal{W}(\mathcal{U}_{D,T}))$ for $x \in \mathcal{H}$.

Example 1.13. (*Twisted Tensor Product*). The above example can be nicely generalized, pointing out the complexity possible in unitary systems having wandering vectors. Let $\mathcal{U}_1$ and $\mathcal{U}_2$ be unitary systems with wandering vectors on Hilbert spaces $\mathcal{H}_1$ and $\mathcal{H}_2$, respectively, and let $\psi_i, \in \mathcal{W}(\mathcal{U}_i), i = 1, 2$. Let $\mathcal{H} = \mathcal{H}_1 \otimes \mathcal{H}_2$. Let $\widetilde{\mathcal{U}}_1 = \mathcal{U}_1 \otimes I$, and let $\widetilde{\mathcal{U}}_2$ be *any* unitary system in $\mathcal{H}$ leaving the subspace $\psi_1 \otimes \mathcal{H}_2$ invariant such that the map
$$U \to U|_{\psi_1 \otimes \mathcal{H}_2}$$

is 1-1 on $\widetilde{\mathcal{U}}_2$ and with
$$\widetilde{\mathcal{U}}_2|_{\psi_1 \otimes \mathcal{H}_2} = (I \otimes \mathcal{U}_2)|_{\psi_1 \otimes \mathcal{H}_2}.$$

Then
$$\mathcal{U} = \widetilde{\mathcal{U}}_1 \widetilde{\mathcal{U}}_2$$

is a unitary system on $\mathcal{H}$. Clearly $\psi_1 \otimes \psi_2 \in \mathcal{W}(\mathcal{U})$. By repeating this procedure one may construct "twisted tensor products" of arbitrary length. The case when $\mathcal{U}_1, \mathcal{U}_2$ are abelian groups is most relevant to wavelet theory, and models the higher-dimensional translation-dilation systems on $\mathbb{R}^n$ which have been studied. In this case $\mathcal{U}_2$ can correspond to the group generated by the translations in the n directions, and $\mathcal{U}_1$ can correspond to an abelian group of dilation unitaries, which can be matrix dilations.

Example 1.14. Without additional structural hypotheses, pathological unitary systems are easily constructed. For instance, if $\{e_n\}_{n=1}^{\infty}$ is an orthonormal basis, let $\mathcal{U}_1 = I$, and for each $n \geq 2$ let U_n be an arbitrary unitary operator with $Ue_1 = e_n$. Then
$$\mathcal{U} = \{U_n : n \in \mathbb{N}\}$$
is a unitary system with e_1 as a complete wandering vector. For certain (likely "most") choices of U_n one will simply have
$$\mathcal{W}(\mathcal{U}) = \{\lambda e_1 : |\lambda| = 1\} \quad \text{and} \quad \mathcal{C}_{e_1}(\mathcal{U}) = \mathbb{C}I.$$

For instance, if U_n is the permutation unitary that interchanges e_1 and e_n and fixes then other basis vectors, then $\mathcal{U}$ will have this property.

CHAPTER 2

Structural Theorems

Let $\mathcal{U}$ be a unitary system in $\mathcal{B}(\mathcal{H})$, and suppose $\mathcal{U}$ contains a subset $\mathcal{U}_0$ which is a group such that $\mathcal{U}\mathcal{U}_0 = \mathcal{U}$. This is the situation for the wavelet theory case

$$\mathcal{U} = \{D^n T^l : n, l \in \mathbb{Z}\},$$

where $\mathcal{U}_0 = \{T^l : l \in \mathbb{Z}\}$. Suppose $\psi \in \mathcal{W}(\mathcal{U})$. Then $\mathcal{U}_0\psi \subseteq \mathcal{W}(\mathcal{U})$ clearly. However, $\mathcal{U}_0$ will not usually be contained in $\mathcal{C}_\psi(\mathcal{U})$. For each $U \in \mathcal{U}_0$, let V_U be the unique unitary in $\mathcal{C}_\psi(\mathcal{U})$ with $V_U\psi = U\psi$ given by Proposition 1.3. Let

$$\kappa_\psi : \mathcal{U}_0 \to \mathbb{U}(\mathcal{C}_\psi(\mathcal{U}))$$

denote the map $\kappa_\psi(U) = V_U, U \in \mathcal{U}_0$.

Theorem 2.1. *With the above notation, $\kappa_\psi(\mathcal{U}_0)$ is a group and κ_ψ is a group anti-isomorphism. The set $\mathcal{U}_0\psi$ is contained in a connected subset of $\mathcal{W}(\mathcal{U})$.*

Proof. Suppose $U_1, U_2 \in \mathcal{U}_0$. Let $S \in \mathcal{U}$. Then

$$
\begin{aligned}
\kappa_\psi(U_2)\kappa_\psi(U_1)S\psi &= \kappa_\psi(U_2)S\kappa_\psi(U_1)\psi \\
&= \kappa_\psi(U_2)SU_1\psi \\
&= SU_1\kappa_\psi(U_2)\psi \\
&= SU_1U_2\psi \\
&= S\kappa_\psi(U_1U_2)\psi \\
&= \kappa_\psi(U_1U_2)S\psi.
\end{aligned}
$$

So $\kappa_\psi(U_2)\kappa_\psi(U_1)$ agrees with $\kappa_\psi(U_1U_2)$ on the orthonormal basis $\mathcal{U}\psi$, and hence they are equal. This shows that $\kappa_\psi(\mathcal{U}_0)$ is a group and that κ_ψ is an anti-homomorphism. If $U \in \mathcal{U}_0$ and $U \neq I$, then $U\psi \neq \psi$ since ψ is wandering for $\mathcal{U}$. Hence $\kappa_\psi(U) \neq I$. So κ_ψ is one-to-one, as required. To show that $\mathcal{U}_0\psi$ is contained in a component, note that the closure in the strong operator topology of the span of $\kappa_\psi(\mathcal{U}_0)$ is the von Neumann algebra $\mathrm{w}^*(\kappa_\psi(\mathcal{U}_0))$, and is contained in $\mathcal{C}_\psi(\mathcal{U})$. The unitary group of a von Neumann algebra is norm connected (c.f. [**14**]). So as in Example 1.10, continuity of the map $V \to V\psi$ from $\mathbb{U}(\mathrm{w}^*(\kappa_\psi(\mathcal{U}_0))) \to \mathcal{H}$ implies that $\mathbb{U}(\mathrm{w}^*(\kappa_\psi(\mathcal{U}_0)))\psi$ is connected in $\mathcal{W}(\mathcal{U})$. This contains $\mathcal{U}_0\psi$ since $U\psi = \kappa_\psi(U)\psi, U \in \mathcal{U}_0$. $\square$

Corollary 2.2. *With the above terminology, if $\mathcal{U}_0$ is nontrivial, then the connected components of $\mathcal{W}(\mathcal{U})$ are all nontrivial.*

Theorem 2.3. *With the above terminology, suppose $\mathcal{U}_0$ is abelian. In this case, if $U \in \mathbb{U}(\mathrm{w}^*(\mathcal{U}_0))$, then $U\mathcal{W}(\mathcal{U}) = \mathcal{W}(\mathcal{U})$. The map κ_ψ extends to a homomorphism of $\mathbb{U}(\mathrm{w}^*(\mathcal{U}_0))$ into $\mathbb{U}(\mathcal{C}_\psi(\mathcal{U}))$.*

Proof. Let $\eta = U\psi$. Let

$$E_\psi = [\mathcal{U}_0\psi] = [\mathrm{w}^*(\mathcal{U}_0)\psi].$$

Then E_ψ reduces $\mathrm{w}^*(\mathcal{U}_0)$. So $UE_\psi = E_\psi$. Suppose $W \in \mathcal{U}$ but $W \notin \mathcal{U}_0$. Then $WV_1 \notin \mathcal{U}_0$ for all $V_1 \in \mathcal{U}_0$, so $WV_1\psi \perp V_2\psi$ for all $V_1, V_2 \in \mathcal{U}_0$. Hence $WE_\psi \perp E_\psi$. More generally, if $W_1, W_2 \in \mathcal{U}$ and $W_1\mathcal{U}_0 \neq W_2\mathcal{U}_0$, then

$$W_1\mathcal{U}_0 \cap W_2\mathcal{U}_0 = \emptyset,$$

so $W_1\mathcal{U}_0\psi \perp W_2\mathcal{U}_0\psi$, and hence $W_1E_\psi \perp W_2E_\psi$. So if $W_1, W_2 \in \mathcal{U}$ and $W_1\mathcal{U}_0 \neq W_2\mathcal{U}_0$, then since $\eta = U\psi \in E_\psi$, we have $W_1\eta \perp W_2\eta$. On the other hand, if $W_1 \neq W_2$ but $W_1\mathcal{U}_0 = W_2\mathcal{U}_0$, then $W_2 = W_1U_1$ for some $U_1 \in \mathcal{U}_0, U_1 \neq I$. Then $U_1\psi \perp \psi$, so $UU_1\psi \perp U\psi$. Since by hypothesis $\mathcal{U}_0$ is abelian, so is $\mathrm{w}^*(\mathcal{U}_0)$, and so $UU_1 = U_1U$. Thus $U_1\eta \perp \eta$. Hence

$$W_2\eta = W_1U_1\eta \perp W_1\eta.$$

We have shown that $\mathcal{U}\eta$ is an orthonormal set. We have

$$[\mathcal{U}_0\eta] = [\mathcal{U}_0U\psi] = [\mathrm{w}^*(\mathcal{U}_0)U\psi] = [\mathrm{w}^*(\mathcal{U}_0)\psi] = E_\psi.$$

So $[\mathcal{U}\eta] = [\mathcal{U}\mathcal{U}_0\eta] = [\mathcal{U}E_\psi] \supseteq [\mathcal{U}\psi] = \mathcal{H}$. Thus $\mathcal{U}\eta$ is complete. So $\eta \in \mathcal{W}(\mathcal{U})$.

Next, for each $U \in \mathbb{U}(\mathrm{w}^*(\mathcal{U}_0))$, let V_U be the unique unitary in $\mathcal{C}_\psi(\mathcal{U})$ for which $V_U\psi = U\psi$ that is given by Proposition 1.3, and define $\kappa_\psi(U) = V_U$. If $U_1, U_2 \in \mathbb{U}(\mathrm{w}^*(\mathcal{U}_0))$, let $S \in \mathcal{U}$ be arbitrary. Then as in Theorem 2.1,

$$\kappa_\psi(U_2)\kappa_\psi(U_1)S\psi = \kappa_\psi(U_2)S\kappa_\psi(U_1)\psi = \kappa_\psi(U_2)SU_1\psi.$$

Since SU_1 is in the strongly closed linear span of $\mathcal{U}$, and since $\kappa_\psi(U_2)$ commutes locally at ψ with each element of $\mathcal{U}$, we have

$$\kappa_\psi(U_2)SU_1\psi = SU_1\kappa_\psi(U_2)\psi = SU_1U_2\psi.$$

Since $U_1, U_2 \in \mathbb{U}(\mathrm{w}^*(\mathcal{U}_0))$ we have $U_1U_2 \in \mathbb{U}(\mathrm{w}^*(\mathcal{U}_0))$. Thus

$$SU_1U_2\psi = S\kappa_\psi(U_1U_2)\psi = \kappa_\psi(U_1U_2)S\psi.$$

So, as in Theorem 2.1, $\kappa_\psi(U_2U_1) = \kappa_\psi(U_1U_2)$ agrees with $\kappa_\psi(U_2)\kappa_\psi(U_1)$ on an orthonormal basis, so they are equal. $\square$

Remarks 2.4. For the special case of the wavelet system $\langle D, T\rangle$ on $L^2(\mathbb{R})$, Theorem 2.3 implies that if U is a unitary operator in $\mathrm{w}^*(T)$, then for any orthogonal wavelet ψ , $U\psi$ is also a wavelet. This can also be deduced by a function-theoretic argument. (c.f. [**8**].)

Problem B. If $\eta \in \mathcal{W}(\mathcal{U})$ and $\eta \in E_\psi := [\mathcal{U}_0\psi]$, is there a unitary operator U in $\mathrm{w}^*(\mathcal{U}_0)$ such that $\eta = U\psi$?

In the abelian case, and in particular in the case of the wavelet system $\langle D, T\rangle$, the answer is yes. (See Corollary 2.17).

Let $\mathcal{U}, \mathcal{U}_0$ be as in Theorem 2.1. Do not assume $\mathcal{U}_0$ is abelian. Let $E_\psi = [\mathcal{U}_0\psi]$, and let $P_\psi = \mathrm{proj}\,(E_\psi)$.

Lemma 2.5. $A \in \mathcal{C}_\psi(\mathcal{U})$ if and only if $(AS - SA)P_\psi = 0$ for all $S \in \mathcal{U}$.

Proof. If $(AS - SA)P_\psi = 0$, then since $P_\psi \psi = \psi$, we have $(AS - SA)\psi = 0$. Conversely, if $(AS - SA)\psi = 0, S \in \mathcal{U}$, then for all $T \in \mathcal{U}_0$,

$$(AS - SA)T\psi = A(ST)\psi - S(AT\psi) = (ST)A\psi - STA\psi = 0.$$

So $(AS - SA)\mathcal{U}_0\psi = 0$, hence $(AS - SA)P_\psi = 0$. $\square$

Lemma 2.6. *If* $\psi, \eta \in \mathcal{W}(\mathcal{U})$, *and if* $V \in \mathbb{U}(\mathcal{C}_\psi(\mathcal{U}))$ *with* $V\psi = \eta$, *then* $P_\eta = V P_\psi V^*$.

Proof. For each $S \in \mathcal{U}_0$ we have $S\eta = SV\psi = VS\psi$. So $[\mathcal{U}_0\eta] = V[\mathcal{U}_0\psi]$. $\square$

Lemma 2.7. *If* $V \in \mathbb{U}(\mathcal{C}_\psi(\mathcal{U}))$, *and if* $V\psi \in E_\psi$, *then*

$$P_{V\psi} = P_\psi, \quad \mathcal{C}_\psi(\mathcal{U}) = \mathcal{C}_{V\psi}(\mathcal{U}), \quad \text{and} \quad P_\psi V = V P_\psi.$$

Proof. For $S \in \mathcal{U}_0$ we have

$$VS\psi = SV\psi \in SP_\psi\mathcal{H} \subseteq P_\psi\mathcal{H}.$$

So $VP_\psi\mathcal{H} \subseteq P_\psi\mathcal{H}$. Let $\eta = V\psi$. Then $P_\eta = V P_\psi V^*$ by Lemma 2.6, so $P_\eta \leq P_\psi$. Let $\{U_n\}$ be a sequence in $\mathcal{U}$ such that $U_n\mathcal{U}_0 \cap U_m\mathcal{U}_0 = \emptyset$, $n \neq m$, and $\cup_n U_n\mathcal{U}_0 = \mathcal{U}$. Then

$$U_n E_\psi \perp U_m E_\psi$$

if $n \neq m$, and $\bigvee_n U_n E_\psi = \mathcal{H}$. The subspace E_η has the same property. So since $E_\eta \subseteq E_\psi$, we must have $E_\eta = E_\psi$. So $P_\eta = P_\psi$. Then Lemma 2.5 implies

$$\mathcal{C}_\psi(\mathcal{U}) = \mathcal{C}_{V\psi}(\mathcal{U}).$$

Since $P_\eta = V P_\psi V^*$, we have $P_\psi V = V P_\psi$. $\square$

Let $\mathcal{C}_\psi^P(\mathcal{U}) = \mathcal{C}_\psi(\mathcal{U}) \cap \{P_\psi\}'$.

Proposition 2.8. $\mathcal{C}_\psi(\mathcal{U})$ *is a left module over* $\mathcal{U}'$ *and a right module over* $\mathcal{C}_\psi^P(\mathcal{U})$. *In particular,* $\mathcal{C}_\psi^P(\mathcal{U})$ *is an algebra.*

Proof. Let $A \in \mathcal{C}_\psi(\mathcal{U})$. If $B \in \mathcal{U}'$, then for $S \in \mathcal{U}$,

$$BAS\psi = BSA\psi = SBA\psi.$$

Hence $BA \in \mathcal{C}_\psi(\mathcal{U})$. Now let $C \in \mathcal{C}_\psi^P(\mathcal{U})$. Then

$$(AC)S\psi = ASC\psi = ASCP_\psi\psi = ASP_\psi C\psi = SAP_\psi C\psi = S(AC)\psi.$$

where the fourth equality is via Lemma 2.5. Thus $AC \in \mathcal{C}_\psi(\mathcal{U})$. $\square$

Theorem 2.9. $\mathcal{C}_\psi^P(\mathcal{U})$ *is a von Neumann algebra.*

Proof. $\mathcal{C}_\psi^P(\mathcal{U})$ is an algebra , and is strongly closed. We must show it is self-adjoint. Suppose $A \in \mathcal{C}_\psi^P(\mathcal{U})$. We must show that $A^*V\psi = VA^*\psi$ for all $V \in \mathcal{U}$. It will suffice to show that

$$\langle A^*V\psi, W\psi \rangle = \langle VA^*\psi, W\psi \rangle$$

for all $V, W \in \mathcal{U}$.

For $U \in \mathcal{U}_0$, write $\alpha_U(A) = \langle A\psi, U\psi \rangle$. Then

$$A\psi = \sum_{U \in \mathcal{U}_0} \alpha_U(A)U\psi.$$

Since P_ψ reduces A, $A^* E_\psi \subseteq E_\psi$. Write $\alpha_U(A^*) = \langle A^*\psi, U\psi\rangle$. Then

$$A^*\psi = \sum_{U \in \mathcal{U}_0} \alpha_U(A^*)U\psi.$$

We have

$$\alpha_U(A^*) = \langle A^*\psi, U\psi\rangle = \langle \psi, AU\psi\rangle = \langle \psi, UA\psi\rangle = \langle U^*\psi, A\psi\rangle = \overline{\alpha_{U^*}(A)}.$$

Now compute:

$$
\begin{aligned}
\langle A^*V\psi, W\psi\rangle = \langle V\psi, AW\psi\rangle &= \langle V\psi, WA\psi\rangle \\
&= \langle V\psi, W \sum_{U \in \mathcal{U}_0} \alpha_U(A)U\psi\rangle \\
&= \langle V\psi, W \sum_{U \in \mathcal{U}_0} \alpha_{U^*}(A)U^*\psi\rangle \\
&= \sum_{U \in \mathcal{U}_0} \alpha_U(A^*)\langle V\psi, WU^*\psi\rangle.
\end{aligned}
$$

We have $WU^*\psi = W\kappa_\psi(U^*)\psi = \kappa_\psi(U^*)W\psi$. So

$$
\begin{aligned}
\langle V\psi, WU^*\psi\rangle &= \langle V\psi, \kappa_\psi(U^*)W\psi\rangle \\
&= \langle \kappa_\psi(U)V\psi, W\psi\rangle \\
&= \langle V\kappa_\psi(U)\psi, W\psi\rangle \\
&= \langle VU\psi, W\psi\rangle \\
&= \langle U\psi, V^*W\psi\rangle.
\end{aligned}
$$

And so

$$
\begin{aligned}
\langle A^*V\psi, W\psi\rangle &= \sum_{U \in \mathcal{U}_0} \alpha_U(A^*)\langle U\psi, V^*W\psi\rangle \\
&= \langle \sum_{U \in \mathcal{U}_0} \alpha_U(A^*)U\psi, V^*W\psi\rangle \\
&= \langle A^*\psi, V^*W\psi\rangle \\
&= \langle VA^*\psi, W\psi\rangle,
\end{aligned}
$$

as required. $\square$

Corollary 2.10. $\mathbb{U}(\mathcal{C}_\psi^P(\mathcal{U}))$ is a group.

Corollary 2.11. $\mathcal{W}(\mathcal{U}) \cap E_\psi$ is connected.

Proof. This follows from Lemma 2.7 and Corollary 2.10. $\square$

Note that if $\mathcal{U}_0$ is abelian, then $\kappa_\psi(U)P_\psi = UP_\psi$ for all $U \in \mathcal{U}_0$. (This can fail if $\mathcal{U}_0$ is nonabelian. For instance, we may have $\mathcal{U} = \mathcal{U}_0$ is a commutative group. Then $P_\psi = I$, and $\mathcal{C}_\psi(\mathcal{U}) = \{\mathcal{U}\}' \not\supseteq \mathcal{U}$.) To see this, note that for commutative $\mathcal{U}_0$, for each $U, V \in \mathcal{U}_0$,

$$\kappa_\psi(U)V\psi = V\kappa_\psi(U)\psi = VU\psi = UV\psi.$$

So since $E_\psi = [\mathcal{U}_0\psi]$, $\kappa_\psi(U)|_{E_\psi} = U|_{E_\psi}$.

Theorem 2.12. Suppose $\mathcal{U}_0$ is abelian. Then $\mathcal{C}_\psi^P(\mathcal{U})$ is abelian.

Proof. Let $A, B \in \mathcal{C}^P_\psi(\mathcal{U})$. Let $U \in \mathcal{U}_0$. Then

$$A\kappa_\psi(U)\psi = AU\psi = UA\psi = UP_\psi A\psi = \kappa_\psi(U)P_\psi A\psi = \kappa_\psi(U)A\psi.$$

Since $A, \kappa_\psi(U) \in \mathcal{C}^P_\psi(\mathcal{U})$, which is an algebra, and since ψ separates $\mathcal{C}^P_\psi(\mathcal{U})$, this shows that

$$A\kappa_\psi(U) = \kappa_\psi(U)A.$$

Since $A\psi \in E_\psi$, we have

$$A\psi = \sum_{U \in \mathcal{U}_0} \alpha_U(A)U\psi,$$

where $\alpha_U(A) = \langle A\psi, U\psi \rangle$. Similarly,

$$B\psi = \sum_{U \in \mathcal{U}_0} \alpha_U(B)U\psi.$$

Let $V \in \mathcal{U}_0$ be arbitrary. Compute

$$
\begin{aligned}
\langle AB\psi, V\psi \rangle &= \sum_{U \in \mathcal{U}_0} \alpha_U(B)\langle AU\psi, V\psi \rangle \\
&= \sum_{U \in \mathcal{U}_0} \alpha_U(B)\langle UA\psi, V\psi \rangle \\
&= \sum_{U \in \mathcal{U}_0} \alpha_U(B)\langle A\psi, U^*V\psi \rangle \\
&= \sum_{U \in \mathcal{U}_0} \alpha_U(B)\alpha_{U^*V}(A).
\end{aligned}
$$

Similarly,

$$
\begin{aligned}
\langle BA\psi, V\psi \rangle &= \sum_{U \in \mathcal{U}_0} \alpha_U(A)\alpha_{U^*V}(B) \\
&= \sum_{W \in \mathcal{U}_0} \alpha_{VW^*}(A)\alpha_W(B) \\
&= \sum_{W \in \mathcal{U}_0} \alpha_W(B)\alpha_{W^*V}(A),
\end{aligned}
$$

where we use the fact that $\mathcal{U}_0$ is abelian. This shows that

$$\langle AB\psi, V\psi \rangle = \langle BA\psi, V\psi \rangle.$$

So $\langle (AB - BA)\psi, V\psi \rangle = 0$, $V \in \mathcal{U}_0$, and since $(AB - BA)\psi \in [\mathcal{U}_0\psi]$, this shows that $(AB - BA)\psi = 0$. So since $AB - BA \in \mathcal{C}^P_\psi(\mathcal{U})$, this implies $AB = BA$, as required. $\square$

The following gives a simple but useful *structural* description of the local commutant for an important special case.

Proposition 2.13. *Let $\mathcal{U}_1$ and $\mathcal{U}_0$ be unitary groups in $\mathcal{B}(\mathcal{H})$, and let $\mathcal{U}$ be the unitary system $\mathcal{U} = \mathcal{U}_1\mathcal{U}_0 = \{UV : U \in \mathcal{U}_1, V \in \mathcal{U}_0\}$. If $\psi \in \mathcal{W}(\mathcal{U})$, then*

$$\mathcal{C}_\psi(\mathcal{U}) = \mathcal{U}'_1 \cap \{\mathcal{U}'_0 + \mathcal{B}(\mathcal{H})P^\perp_\psi\}.$$

Proof. First note that

$$\mathcal{U}'_0 + \mathcal{B}(\mathcal{H})P^\perp_\psi = \{A \in \mathcal{B}(\mathcal{H}) : AP_\psi \in \mathcal{U}'_0 P_\psi\}.$$

Indeed, if $AP_\psi \in \mathcal{U}_0' P_\psi$ then $AP_\psi = RP_\psi$ for some $R \in \mathcal{U}_0'$, so

$$A = R + (A - R)P_\psi^\perp \in \mathcal{U}_0' + \mathcal{B}(\mathcal{H})P_\psi^\perp.$$

Now suppose $A \in \mathcal{C}_\psi(\mathcal{U})$. Then $A \in \mathcal{U}_1'$ by Lemma 1.1 (v). Also, for any $U \in \mathcal{U}_0$ we have $UAP_\psi = AUP_\psi = AP_\psi U$. Thus $AP_\psi \in \mathcal{U}_0'$, and so $AP_\psi \in \mathcal{U}_0' P_\psi$. Thus

$$A \in \mathcal{U}_0' + \mathcal{B}(\mathcal{H})P_\psi^\perp$$

by the above paragraph. Conversely, if $A \in \mathcal{U}_1'$ and $A \in \mathcal{U}_0' + \mathcal{B}(\mathcal{H})P_\psi^\perp$, then $A = B + CP_\psi^\perp$ for some $B \in \mathcal{U}_0'$. For $U \in \mathcal{U}_0$ we have $P_\psi^\perp U\psi = 0$. So for $V \in \mathcal{U}_1$, $U \in \mathcal{U}_0$ we have

$$AVU\psi = VAU\psi = V(B + CP_\psi^\perp)U\psi = VBU\psi = VUB\psi = VUA\psi.$$

That is, $A \in \mathcal{C}_\psi(\mathcal{U})$. $\square$

Lemma 2.14. *Let $\mathcal{U}_0, \mathcal{U}_1$ and $\mathcal{U}$ be as in Proposition 2.13 with $\mathcal{U}_1 \cap \mathcal{U}_0 = \{I\}$. Let $\psi \in \mathcal{W}(\mathcal{U})$. (Note that $\mathcal{U}$ must be countable.) Then for all $A \in \mathcal{B}(\mathcal{H})$, the sum $\sum_{U \in \mathcal{U}_1} UP_\psi AP_\psi U^*$ converges in the strong operator topology to an element of $\mathcal{U}_1'$.*

Proof. The operators

$$\{UP_\psi AP_\psi U^{-1} : U \in \mathcal{U}_1\}$$

are uniformly bounded and have mutually orthogonal ranges and mutually orthogonal supports. So any enumeration of $\mathcal{U}_1$ leads to a convergent sum, and the limit is independent of the enumeration.

If $V \in \mathcal{U}_1$, then

$$V \sum_{U \in \mathcal{U}_1} UP_\psi AP_\psi U^{-1} = (\sum_{U \in \mathcal{U}_1} (VU)P_\psi AP_\psi (VU)^{-1})V.$$

Since VU is a generic element of the group $\mathcal{U}_1$, this shows that V commutes with the sum, as required. $\square$

Lemma 2.15. *Let $\mathcal{U}_0, \mathcal{U}_1, \mathcal{U}, \psi$ be as above. Then*

$$\mathcal{C}_\psi^P(\mathcal{U}) = \{\sum_{U \in \mathcal{U}_1} UP_\psi SP_\psi U^* : S \in \mathcal{U}_0'\}.$$

Proof. Suppose $S \in \mathcal{U}_0'$. Then $B := \sum_{U \in \mathcal{U}_1} UP_\psi SP_\psi U^*$ is in $\mathcal{U}_1'$ by Lemma 2.14. If $U \in \mathcal{U}_1$ and $U \neq I$ then $P_\psi UP_\psi = 0$, so

$$P_\psi B = BP_\psi = P_\psi SP_\psi.$$

Since $P_\psi \in \mathcal{U}_0'$ we have $P_\psi SP_\psi \in \mathcal{U}_0' P_\psi$. Thus $B \in \mathcal{C}_\psi^P(\mathcal{U})$, using Proposition 2.13 and the first line of its proof.

Conversely, suppose $A \in \mathcal{C}_\psi^P(\mathcal{U})$. Then $A \in \mathcal{U}_1'$ and $AP_\psi = P_\psi A$. Also, $AP_\psi \in \mathcal{U}_0' P_\psi$. Choose $S \in \mathcal{U}_0'$ such that $AP_\psi = SP_\psi$. Let $B = \sum_{U \in \mathcal{U}_1} UP_\psi SP_\psi U^*$. Then $B \in \mathcal{C}_\psi^P(\mathcal{U})$ by the first paragraph. We have

$$BP_\psi = P_\psi SP_\psi = P_\psi AP_\psi = AP_\psi.$$

So $B\psi = A\psi$. Thus $A = B$ by Lemma 1.1 (i). $\square$

Theorem 2.16. *If* $\mathcal{U} = \mathcal{U}_1\mathcal{U}_0$ *with* $\mathcal{U}_1, \mathcal{U}_0$ *groups, with* $\mathcal{U}_1 \cap \mathcal{U}_0 = \{I\}$, *and with* $\mathcal{U}_0$ *abelian, then*

1. $\mathcal{C}_\psi^P(\mathcal{U}) = \mathrm{w}^*(\kappa_\psi(\mathcal{U}_0))$;
2. $\mathrm{w}^*(\mathcal{U}_0)P_\psi = \mathcal{C}_\psi^P(\mathcal{U})P_\psi$;
3. $\mathcal{C}_\psi^P(\mathcal{U})$ *is* $*$-*isomorphic to* $\mathrm{w}^*(\mathcal{U}_0)|_{E_\psi}$;
4. κ_ψ *extends to a* $*$-*homomorphism from* $\mathrm{w}^*(\mathcal{U}_0)$ *onto* $\mathcal{C}_\psi^P(\mathcal{U})$.

Proof. For $U \in \mathcal{U}_1$, if $U \neq I$ then $P_\psi U P_\psi = 0$, so by Lemma 2.15,

$$\mathcal{C}_\psi^P(\mathcal{U})P_\psi = P_\psi \mathcal{U}_0' P_\psi.$$

Since $P_\psi \in \mathcal{U}_0'$, this is a von Neumann algebra. Theorem 2.12 implies it is abelian. Note that $\kappa_\psi(\mathcal{U}_0) \subseteq \mathcal{C}_\psi^P(\mathcal{U})$ by Lemma 2.7. So $\mathrm{w}^*(\kappa_\psi(\mathcal{U}_0)) \subseteq \mathcal{C}_\psi^P(\mathcal{U})$, and so

$$\mathcal{C}_\psi^P(\mathcal{U})P_\psi \supseteq \mathrm{w}^*(\kappa_\psi(\mathcal{U}_0))P_\psi.$$

Since $P_\psi \in (\mathrm{w}^*(\kappa_\psi(\mathcal{U}_0)))'$ and $\mathrm{w}^*(\kappa_\psi(\mathcal{U}_0))P_\psi$ is a von Neumann algebra, $\mathrm{w}^*(\kappa_\psi(\mathcal{U}_0))P_\psi$ is w$*$-closed. The vector ψ is cyclic for $\mathrm{w}^*(\kappa_\psi(\mathcal{U}_0))|_{P_\psi \mathcal{H}}$, so this compression algebra is a m.a.s.a. So since $\mathcal{C}_\psi^P(\mathcal{U})P_\psi$ is abelian, we must have

$$\mathcal{C}_\psi^P(\mathcal{U})P_\psi = \mathrm{w}^*(\kappa_\psi(\mathcal{U}_0))P_\psi.$$

Similarly, $P_\psi \in (\mathrm{w}^*(\mathcal{U}_0))'$, so $\mathrm{w}^*(\mathcal{U}_0)P_\psi$ is w$*$-closed and abelian. For $U \in \mathcal{U}_0$ we have $U P_\psi = \kappa_\psi(U)P_\psi$, so $\mathrm{w}^*(\mathcal{U}_0)P_\psi \supseteq \{\kappa_\psi(U)P_\psi : U \in \mathcal{U}_0\}$, and hence

$$\mathrm{w}^*(\mathcal{U}_0)P_\psi \supseteq \mathrm{w}^*(\kappa_\psi(\mathcal{U}_0))P_\psi.$$

The reverse inequality is similar. Thus $\mathrm{w}^*(\mathcal{U}_0)P_\psi = \mathrm{w}^*(\kappa_\psi(\mathcal{U}_0))P_\psi$. Hence $\mathrm{w}^*(\mathcal{U}_0)P_\psi = \mathcal{C}_\psi^P(\mathcal{U})P_\psi$. Item (ii) is proven.

Let θ be the map defined in Lemma 2.14. That is

$$\theta := \sum_{U \in \mathcal{U}_1} U P_\psi A P_\psi U^*, \ A \in \mathcal{B}(\mathcal{H}).$$

Then θ is linear, and $\theta(A^*) = (\theta(A))^*$, $A \in \mathcal{B}(\mathcal{H})$. To prove item (i) it suffices to prove " $\subseteq$ ". Suppose $B \in \mathcal{C}_\psi^P(\mathcal{U})$. Then

$$P_\psi B P_\psi = P_\psi A P_\psi$$

for some $A \in \mathrm{w}^*(\kappa_\psi(\mathcal{U}_0))$. By the second paragraph of the proof of Lemma 2.15, $A = \theta(A)$ and $B = \theta(B)$. So

$$B = A \in \mathrm{w}^*(\kappa_\psi(\mathcal{U}_0)),$$

as required.

For $V \in \mathcal{U}_0, V P_\psi = \kappa_\psi(V)P_\psi$. As above, $\theta(\kappa_\psi(V)) = \kappa_\psi(V)$. Hence $\theta(V) = \kappa_\psi(V)$. If $V \in \mathbb{U}(\mathrm{w}^*(\mathcal{U}_0))$, then

$$\theta(V) \in \mathcal{C}_\psi^P(\mathcal{U}) \subset \mathcal{C}_\psi(\mathcal{U})$$

by Lemma 2.15, and also $\kappa_\psi(V) \in \mathcal{C}_\psi(\mathcal{U})$. Since V is unitary and commutes with P_ψ, $\theta(V)$ is unitary. Also,

$$\theta(V)P_\psi = V P_\psi = \kappa_\psi(V)P_\psi.$$

So by the uniqueness part of Proposition 1.3, $\theta(V) = \kappa_\psi(V)$. Hence $\theta(V) = \kappa_\psi(V)$ where κ_ψ was previously defined, so θ extends κ_ψ. Since $P_\psi \in \mathcal{U}_0'$, θ is multiplicative on $\mathrm{w}^*(\mathcal{U}_0)$. Thus $\theta|_{\mathrm{w}^*(\mathcal{U}_0)}$ is a $*$-homomorphism. Since $\theta(A) = 0$

iff $P_\psi A P_\psi = 0$, θ is 1-1 on $\mathrm{w}^*(\mathcal{U}_0)P_\psi$. So it induces a $*$-isomorphism between $P_\psi \mathrm{w}^*(\mathcal{U}_0)|_{E_\psi}$ and $\mathcal{C}_\psi^P(\mathcal{U})$. $\square$

Corollary 2.17. *With the hypotheses of Theorem 2.16, if $\psi, \eta \in \mathcal{W}(\mathcal{U})$ with $\eta \in E_\psi$, then there is a unitary $V \in \mathrm{w}^*(\mathcal{U}_0)$ with $\eta = V\psi$.*

Proof. Let $W \in \mathbb{U}(\mathcal{C}_\psi^P(\mathcal{U}))$ with $W\psi = \eta$. By Lemma 2.7, $W \in \mathcal{C}_\psi^P(\mathcal{U})$. Write $W = e^{iA}$ with $A \in \mathcal{C}_\psi^P(\mathcal{U})$ and $A = A^*$. By Theorem 2.16 there exists $B = B^* \in \mathrm{w}^*(\mathcal{U}_0)$ with $A = \kappa_\psi(B)$. Let $V = e^{iB}$. Then V is unitary, and $\kappa_\psi(V) = W$. Since

$$\kappa_\psi(V) = \sum_{U \in \mathcal{U}_1} U P_\psi V P_\psi U^*,$$

we have $WP_\psi = VP_\psi$, so $\eta = W\psi = V\psi$, as required. $\square$

Remark 2.18. If $\mathcal{U}$ is a unitary system with $\mathcal{W}(\mathcal{U}) \neq \emptyset$, let us use the term *wandering vector multiplier* for $\mathcal{U}$ to denote a unitary operator V with the property that $V\mathcal{W}(\mathcal{U}) \subseteq \mathcal{W}(\mathcal{U})$. (That is, we require that $V\psi \in \mathcal{W}(\mathcal{U})$ for *all* $\psi \in \mathcal{W}(\mathcal{U})$, not simply for a *specific* ψ as with unitaries in $\mathcal{C}_\psi(\mathcal{U})$, and we do *not* require that V is in $\mathcal{C}_\psi(\mathcal{U})$ for any ψ.) Every unitary in $\mathcal{U}'$ is a w.v. multiplier since $\mathcal{U}' \subseteq \mathcal{C}_\psi(\mathcal{U})$ for every ψ. In the special case when $\mathcal{U} = \mathcal{U}_1\mathcal{U}_0$ with $\mathcal{U}_1, \mathcal{U}_0$ groups and with $\mathcal{U}_0$ abelian, by Theorem 2.3 (see also Remark 2.4) every unitary $V \in \mathrm{w}^*(\mathcal{U}_0)$ is a w.v. multiplier. In fact, Corollary 2.17 states that for a given $\psi \in \mathcal{W}(\mathcal{U})$ every wandering vector in E_ψ can be attained by acting on ψ by a unitary "multiplier" of this form. So we have a natural problem.

Problem C. (Factorization problem) *Suppose $\mathcal{U}$ is a unitary system of the form $\mathcal{U} = \mathcal{U}_1\mathcal{U}_0$, with $\mathcal{U}_1, \mathcal{U}_0$ groups, $\mathcal{U}_0$ abelian. Is every wandering vector multiplier for $\mathcal{U}$ of the form $V = V_1V_0$ with V_1 a unitary in $\mathcal{U}'$ and V_0 a unitary in $\mathrm{w}^*(\mathcal{U}_0)$?*

Note that these two types of unitaries commute, so if the answer is yes, then the set of wandering vector multipliers will be an abelian group if $\mathcal{U}'$ is abelian. For the special case of $\langle D, T \rangle$ this question has particular interest. (See Remark 3.7.) Let us refer to unitaries in $\mathcal{U}'$ as w.v. multipliers of the *first type*, and to unitaries in $\mathrm{w}^*(\mathcal{U}_0)$ as w.v. multipliers of the *second type*. In the case where $\mathcal{U}$ is a wavelet system, we refer to w.v. multipliers as wavelet multipliers.

CHAPTER 3

The Wavelet System $\langle D, T \rangle$

Let T and D be the translation and dilation unitary operators on $L^2(\mathbb{R})$ described in the introduction given by

$$(Tf)(t) = f(t-1) \quad \text{and} \quad (Df)(t) = \sqrt{2}f(2t).$$

For $f \in L^2(\mathbb{R})$ we have

$$(TDf)(t) = T(\sqrt{2}f(2t)) = \sqrt{2}f(2(t-1)) = \sqrt{2}f(2t-2) = (DT^2 f)(t),$$

so $TD = DT^2$. As earlier, let

$$\langle D, T \rangle := \mathcal{U}_{D,T} := \{D^n T^l : n, l \in \mathbb{Z}\}.$$

We will abbreviate

$$\mathcal{W}(D, T) := \mathcal{W}(\mathcal{U}_{D,T})$$

for the set of wandering vectors for $\mathcal{U}_{D,T}$. These are the orthogonal wavelets. We will also abbreviate

$$\mathcal{C}_\psi(D, T) := \mathcal{C}_\psi(\mathcal{U}_{D,T}).$$

We first recapitulate some of the results of Section 1 as applied to the system $\langle D, T \rangle$.

Lemma 3.1. *Let ψ be any fixed wavelet for $\langle D, T \rangle$. Then*

1. $\mathcal{W}(D, T) = \mathbb{U}(\mathcal{C}_\psi(D, T))\psi$. *The mapping*

$$\psi \to U\psi \quad \text{from} \quad \mathbb{U}(\mathcal{C}_\psi(D, T)) \to \mathcal{W}(D, T)$$

   is one-to-one and onto.
2. $T \notin \mathcal{C}_\psi(D, T)$ *and* $D \notin \mathcal{C}_\psi(D, T)$.
3. $\mathcal{C}_\psi(D, T) \subseteq \{D\}'$.
4. *If* $\eta \in \mathcal{W}(D, T)$, *let* $V \in \mathcal{C}_\psi(D, T)$ *with* $V\psi = \eta$. *Then*

$$\mathcal{C}_\eta(D, T) = \mathcal{C}_\psi(D, T)V^*.$$

Proof. Item (i) is a special case of Proposition 1.3, and items (ii), (iii) and (iv) are special cases of Lemma 1.1 parts (iv),(v) and (vi), respectively. $\square$

If ψ is an orthogonal wavelet, then $T^n\psi$ is an orthogonal wavelet for all $n \in \mathbb{Z}$, even though $T \notin \mathcal{C}_\psi(D, T)$. This is simply because $\mathcal{U}_{D,T} \cdot T^n = \mathcal{U}_{D,T}$. (See the remark before Theorem 2.1.) More generally, if V is any unitary in $\mathrm{w}^*(T)$, then $V\mathcal{W}(D, T) = \mathcal{W}(D, T)$. As in Chapter 2, let $V_\psi = \kappa_\psi(T)$ denote the unique unitary in $\mathcal{C}_\psi(D, T)$ with $V_\psi\psi = T\psi$. Then for each $n \in \mathbb{Z}$, V_ψ^n is the unique unitary in $\mathcal{C}_\psi(D, T)$ with $V_\psi^n\psi = T^n\psi$. In particular, $\mathcal{C}_\psi(D, T)$ contains the group generated by V_ψ. The homomorphism

$$\kappa_\psi : \mathrm{Group}(T) \to \mathcal{C}_\psi(D, T)$$

21

extends uniquely to a homomorphism of

$$\mathbb{U}(\mathrm{w}^*(T)) \to \mathbb{U}(\mathcal{C}_\psi(D, T))$$

with the property that for each $V \in \mathrm{w}^*(T)$, $\kappa_\psi(V)\psi = V\psi$. All this follows from Theorem 2.1 and Theorem 2.3. See also Remark 2.4.

For $\beta \in \mathbb{R}$, let T_β denote the unitary of translation by β :

$$(T_\beta f)(t) = f(t - \beta), \ t \in L^2(\mathbb{R}).$$

Lemma 3.2. *Let* $n \in \mathbb{Z}$ *and* $\beta \in \mathbb{R}$. *Then*

$$D^n T_\beta = T_{2^{-n}\beta} D^n \text{and} \ T_\beta D^n = D^n T_{2^n \beta}.$$

Proof. We have

$$
\begin{aligned}
(D^n T_\beta)(t) &= D^n f(t - \beta) \\
&= (\sqrt{2})^n f(2^n t - \beta),
\end{aligned}
$$

and

$$
\begin{aligned}
(T_{2^{-n}\beta} D^n f)(t) &= T_{2^{-n}\beta}(\sqrt{2})^n f(2^n t) \\
&= (\sqrt{2})^n (f(2^n(t - 2^{-n}\beta))) \\
&= (\sqrt{2})^n f(2^n t - \beta).
\end{aligned}
$$

This establishes the first identity. For the second, replace n and β with $-n$ and $-\beta$ and take adjoints. $\square$

Remark 3.3. If $f \in L^2(\mathbb{R})$ is real valued, then clearly gf is real valued for all $g \in \mathrm{Group}(D, T)$. Also, if an orthogonal wavelet η for $\langle D, T \rangle$ is real valued then η, considered as an element of real $L^2(\mathbb{R})$, is a real wavelet because the real span of $(\mathcal{U}_{D,T})\eta$ is dense in the real L^2-space and the elements of $(\mathcal{U}_{D,T})\eta$ remain orthonormal. Conversely, if η is a real wavelet for $\langle D, T \rangle$, then considered as an element of complex $L^2(\mathbb{R})$ it is also a complex wavelet. So Lemma 3.1 (i) gives a way of parameterizing all real wavelets as well.

Proposition 3.4. *Each (real or complex) path-connected component of* $\mathcal{W}(D, T)$ *in (real or complex)* $L^2(\mathbb{R})$ *is nontrivial.*

Proof. The complex case is a special case of Corollary 2.2, with $\mathcal{U}_0 = \{T^n : n \in \mathbb{Z}\}$. The real case is more subtle since it is special to $\langle D, T \rangle$. Let ψ be a real wavelet. Regard ψ as a wavelet in complex $L^2(\mathbb{R})$. For $0 \le t \le 1$, let

$$W_t := \exp(it\mathrm{Im}(V_\psi)) = \exp(t(\frac{V_\psi - V_\psi^*}{2})).$$

Then W_t is unitary and is contained in $\mathrm{w}^*(V_\psi) \subseteq \mathcal{C}_\psi(D, T)$. Note that the matrix coordinate elements of V_ψ with respect to the wavelet basis $\{D^n T^l \psi : (n, l) \in \mathbb{Z}^2\}$ are all real, hence $W_t = \exp(t(\frac{V_\psi - V_\psi^*}{2}))$ has real coordinates with respect to this basis. So since ψ is a real valued function $W_t \psi$ is real-valued. Thus the path

$$\alpha(t) := W_t \psi, \ 0 \le t \le 1,$$

consists of real wavelets. $\square$

Next we will characterize the *commutant* of $\{D, T\}$ using the Fourier-Plancherel transformation. This is simple, but is apparently new. It will be the first step in our (partial) analysis of $\mathcal{C}_\psi(D, T)$, and hence $\mathcal{W}(D, T)$, beyond the theory accomplished abstractly in Sections 1 and 2.

Lemma 3.2 shows that $DT_\alpha D^{-1} = T_{\frac{\alpha}{2}}, \alpha \in \mathbb{R}$. This implies that Group(D, T) contains the abelian subgroup $\{T_\alpha : \alpha \text{ is dyadic}\}$. It is easy to see that $\chi_{[0,1]}$ is a cyclic vector for the linear span of these dyadic translations. It follows that the closure of this linear span in the strong operator topology is a maximal abelian von Neumann subalgebra of $L^2(\mathbb{R})$. (a m.a.s.a.). Denote this by $\mathcal{A}_T$. Then the commutant $\{D, T\}'$ is contained in $\mathcal{A}_T$. This proves that $\{D, T\}'$ is abelian.

Let $\mathcal{F}$ be the Fourier-Plancherel transform on $\mathcal{H} = L^2(\mathbb{R})$. Then $\mathcal{F}$ is a unitary transformation in $\mathcal{B}(\mathcal{H})$. If $f, g \in L^1(\mathbb{R}) \cap \mathcal{H}$ then

$$(\mathcal{F}f)(s) := \frac{1}{\sqrt{2\pi}} \int_\mathbb{R} e^{-ist} f(t) dt := \widehat{f}(s),$$

and

$$(\mathcal{F}^{-1}g)(t) := \frac{1}{\sqrt{2\pi}} \int_\mathbb{R} e^{ist} g(s) ds.$$

Let α be an arbitrary real number. Then

$$
\begin{aligned}
(\mathcal{F}T_\alpha f)(s) &= \frac{1}{\sqrt{2\pi}} \int_\mathbb{R} e^{-ist} f(t - \alpha) dt \\
&= e^{-is\alpha} (\mathcal{F}f)(s).
\end{aligned}
$$

So $\mathcal{F}T_\alpha \mathcal{F}^{-1} g = e^{-is\alpha} g$. For $A \in \mathcal{B}(\mathcal{H})$ let $\widehat{A}$ denote $\mathcal{F}A\mathcal{F}^{-1}$. Thus $\widehat{T}_\alpha = M_{e^{-i\alpha s}}$. (For $h \in L^\infty$ we use M_h to denote the multiplication operator $f \to hf, f \in L^2$.) Since

$$\{M_{e^{-i\alpha s}} : \alpha \in \mathbb{R}\}$$

generates the m.a.s.a.

$$\mathcal{D}(\mathbb{R}) = \{M_h : h \in L^\infty(\mathbb{R})\}$$

as a von Neumann algebra, we have

$$\mathcal{F}\mathcal{A}_T \mathcal{F}^{-1} = \mathcal{D}(\mathbb{R}).$$

Similarly,

$$
\begin{aligned}
(\mathcal{F}D^n f)(s) &= \frac{1}{\sqrt{2\pi}} \int_\mathbb{R} e^{-ist} (\sqrt{2})^n f(2^n t) dt \\
&= (\sqrt{2})^{-n} \cdot \frac{1}{\sqrt{2\pi}} \int_\mathbb{R} e^{-i2^{-n}st} f(t) dt \\
&= (\sqrt{2})^{-n} (\mathcal{F}f)(2^{-n}s) = (D^{-n} \mathcal{F}f)(s).
\end{aligned}
$$

So $\widehat{D}^n = D^{-n} = D^{*n}$. Therefore, $\widehat{D} = D^{-1} = D^*$.

We have

$$\mathcal{F}\{D, T\}' \mathcal{F}^{-1} = \{\widehat{D}, \widehat{T}\}'$$

and

$$\mathcal{F}\mathbb{U}(\{D, T\}')\mathcal{F}^{-1} = \mathbb{U}(\{\widehat{D}, \widehat{T}\}').$$

Theorem 3.5.

$$\{\widehat{D}, \widehat{T}\}' \;=\; \{M_h : h \in L^\infty(\mathbb{R}) \text{ and } h(s) = h(2s) \text{ a.e. } \}, \text{ and}$$
$$\mathbb{U}(\{\widehat{D}, \widehat{T}\}') \;=\; \{M_h : |h(s)| = 1 \text{ and } h(s) = h(2s) \text{ a.e. } \}.$$

Proof. Since $\widehat{D} = D^*$ and D is unitary, it is clear that $M_h \in \{\widehat{D}, \widehat{T}\}'$ if and only if M_h commutes with D. So let $g \in L^2(\mathbb{R})$ be arbitrary. Then (a.e.) we have

$$(M_h D g)(s) = h(s)(\sqrt{2} g(2s)), \text{ and}$$
$$(D M_h g)(s) = D(h(s) g(s)) = \sqrt{2} h(2s) g(2s).$$

Since these must be equal a.e. for arbitrary g, we must have $h(s) = h(2s)$ a.e. The unimodular condition is necessary and sufficient for M_h to be unitary. $\square$

Remark 3.6. (An algorithm). Let $E = [-2, -1) \cup [1, 2)$, and for $n \in \mathbb{Z}$ let $E_n = \{2^n x : x \in E\}$. Observe that the sets E_n are disjoint and have union $\mathbb{R} \setminus \{0\}$. So if g is any uniformly bounded function on E, then g extends uniquely (a.e.) to a function $\widetilde{g} \in L^\infty(\mathbb{R})$ satisfying $\widetilde{g}(s) = \widetilde{g}(2s), s \in \mathbb{R}$, by setting $\widetilde{g}(2^n s) = g(s), s \in E, n \in \mathbb{Z}$, and $\widetilde{g}(0) = 0$. We have $\|\widetilde{g}\|_\infty = \|g\|_\infty$. Conversely, if h is any function satisfying $h(s) = h(2s)$ a.e., then h is uniquely (a.e.) determined by its restriction to E. This 1-1 mapping $g \to M_{\widetilde{g}}$ from $L^\infty(E)$ onto $\{\widehat{D}, \widehat{T}\}'$ is isometric. It is a $*$-isomorphism when one regards $L^\infty(E)$ as a von Neumann algebra. We will refer to a function h satisfying $h(s) = h(2s)$ a.e. as a *2-dilation periodic function*. This gives a concrete algorithm for computing a large class of wavelets from a given one:

> Given ψ, let $\widehat{\psi} = \mathcal{F}(\psi)$, choose a real-valued function $h \in L^\infty(E)$ arbitrarily, let $g = exp(ih)$, extend to a 2-dilation periodic function $\widetilde{g}$ as above, and compute $\psi_{\widetilde{g}} = \mathcal{F}^{-1}(\widetilde{g}\widehat{\psi})$.

In the description above, the set E could clearly be replaced with $[-2\pi, -\pi) \cup [\pi, 2\pi)$, or with any other "dyadic" set $[-2a, a) \cup [a, 2a)$ for some $a > 0$.

Remark 3.7. (Wavelet multipliers). In the remark above, a concrete description as multiplication operators is given for the group of wavelet multipliers of the first type, via the Fourier transform. It is easily seen, since $\widehat{T} = M_{e^{-is}}$, that $(\mathrm{w}^*(T))\widehat{} := \{\widehat{A} : A \in \mathrm{w}^*(T)\} = \{M_f : f \in L^\infty(\mathbb{R}) \text{ and} f \text{ is } 2\pi\text{-periodic}\}$. So this gives a characterization of wavelet multipliers of the second type, via the Fourier transform, also as multiplication operators. So Problem C has the following function-theoretic subproblem.

Problem C'. *If h is a unimodular function in $L^\infty(\mathbb{R})$ with the property that $\mathcal{F}^{-1}(h\mathcal{F}(\psi))$ is a wavelet for every wavelet ψ, does h necessarily factor $f = f_1 f_2$, where f_1 is 2-dilation-periodic and f_2 is 2π-translation periodic?*

Theorem 3.5, and the algorithm in Remark 3.6 is apparently new to wavelet theory, although it is simple in nature. We have learned that it was also recently obtained, completely independently, in [**15**], for a different type of study.

The fact that 2π-periodic unimodular functions "multiply" wavelets in the sense of Remark 3.7 is not new. For instance a proof is contained in [**8**], in which it is noted that two orthogonal wavelets are associated with the same multiresolution analysis iff they are related by a unimodular 2π-periodic function in this way.

Remark 3.8. Since the unitary group of a von Neumann algebra is connected (in this case, more simply, the set $\{g \in L^\infty(E) : |g(s)| = 1 \text{ a.e.}\}$ is connected) it follows that for fixed ψ the set

$$\{\psi_{\tilde{g}} : g \in L^\infty(E), |g(s)| = 1 \text{ a.e.}\}$$

is a norm-path-connected set of orthogonal wavelets. This gives another proof distinct from Proposition 3.4 that each connected component of $\mathcal{W}(D, T)$ is non-trivial. In Theorem 4.4 we show that V_ψ is *not* contained in $\{D, T\}'$ except for special cases. Thus the types of paths implied by Corollary 2.2 and Theorem 3.5 are necessarily different.

Now let $\psi \in \mathcal{W}(D, T)$ and let P_ψ denote the orthogonal projection onto the *translation space*

$$E_\psi = \overline{\text{span}}\{T^l \psi : l \in \mathbb{Z}\}.$$

(These are the special cases of the E_ψ and P_ψ from Section 2.) Then P_ψ reduces T. Also, since $\{T^l \psi : l \in \mathbb{Z}\}$ is an orthonormal basis for $P_\psi \mathcal{H}$, it follows that the subspaces $\{D^n P_\psi \mathcal{H} : n \in \mathbb{Z}\}$ are mutually orthogonal and span $\mathcal{H}$. So E_ψ is an infinite dimensional wandering subspace for D. As in Section 2 let

$$\mathcal{C}_\psi^P(D, T) = \mathcal{C}_\psi(D, T) \cap \{P_\psi\}'.$$

Recapitulating some results of sections 1 and 2 applied to $\langle D, T \rangle$ we have:

Theorem 3.9. *With the above notation*

1. $A \in \mathcal{C}_\psi(D, T)$ *if and only if*

$$AD^n T^l P_\psi = D^n T^l A P_\psi$$

for all $(n, l) \in \mathbb{Z}^2$.
2. *If* ψ *and* η *are orthogonal wavelets, and if* $V \in \mathbb{U}(\mathcal{C}_\psi(D, T))$ *with* $V\psi = \eta$, *then* $P_\eta = V P_\psi V^*$.
3. *If* $V \in \mathbb{U}(\mathcal{C}_\psi^P(D, T))$ *then* $P_{V\psi} = P_\psi$.
4. *If* $V \in \mathbb{U}(\mathcal{C}_\psi(D, T))$, *and if* $V\psi \in E_\psi$, *then* $V \in \mathcal{C}_\psi^P(D, T)$.
5. *If* ψ *and* η *are wavelets which lie in the same translation space, then*

$$E_\psi = E_\eta, \ \mathcal{C}_\psi(D, T) = \mathcal{C}_\eta(D, T)$$

and

$$\mathcal{C}_\psi^P(D, T) = \mathcal{C}_\eta^P(D, T).$$

6. $\mathcal{C}_\psi(D, T)$ *is a left module over* $\{D, T\}'$ *and a right module over* $\mathcal{C}_\psi^P(D, T)$.
7. *If* ψ *is a wavelet, then*

$$\mathcal{C}_\psi(D, T) = \{D\}' \cap (\{T\}' + \mathcal{B}(\mathcal{H}) P_\psi^\perp).$$

8. $\mathcal{C}_\psi^P(D, T) = \text{w}^*(V_\psi)$, *where* $V_\psi = \kappa_\psi(T)$.
9. $\mathcal{W}(D, T) \cap E_\psi$ *is connected.*

Proof. These follow as special cases of items 2.5 $\to$ 2.8 and 2.11, 2.13, 2.16. $\square$

We next show that if the Fourier transform of a wavelet ψ has full support, then nonscalar unitaries in $\{D, T\}'$ always map ψ out of its translation space. So

$$\mathbb{U}(\{D, T\}') \quad \text{and} \quad \mathbb{U}(\mathcal{C}_\psi^P(D, T))$$

act "disjointly" on ψ in this case. We need a lemma.

Lemma 3.10. *Let f be a function in $L^\infty(\mathbb{R})$. Assume that f is 2π-periodic and 2-dilation-periodic. Then f is a constant (a.e.) function.*

Proof. If f satisfies the hypothesis so do $\operatorname{Re} f$ and $\operatorname{Im} f$. Hence it will suffice to assume that f is real-valued. Assume that f is not constant (a.e.). Then for some real number c both sets

$$E := \{ t \in (-\pi, \pi) : f(t) > c \}$$

and

$$F := \{ t \in (-\pi, \pi) : f(t) < c \}$$

have positive Lebesgue measure. Either $0 < m(E) \leq \pi$ or $0 < m(F) \leq \pi$. Assume the former. The proof for the latter is similar. Since E has positive Lebesgue measure it contains points of *density* ([**22**], p.261). That is, there are points $d_0 \in E$ such that

$$\lim_{h \to 0^+} \frac{m(E \cap [d_0 - h, d_0 + h])}{2h} = 1.$$

Fix such a point d_0, and let n be large enough so that $(d_0 - \frac{\pi}{2^n}, d_0 + \frac{\pi}{2^n}) \subseteq (-\pi, \pi)$ and also that

$$m(E \cap (d_0 - \frac{\pi}{2^n}, d_0 + \frac{\pi}{2^n})) > \frac{3}{4} \cdot 2 \cdot \frac{\pi}{2^n}.$$

Since f is 2π-periodic, we have

$$m(\{ t \in (2^n d_0 - \pi, 2^n d_0 + \pi) : f(t) > c \}) = m(E) \leq \pi.$$

However, since f is 2-dilation-periodic we also have

$$\begin{aligned}
m(\{ t \in (2^n d_0 - \pi, 2^n d_0 + \pi) \quad &: \quad f(t) > c \}) \\
&= \quad 2^n m(\{ t \in (d_0 - \frac{\pi}{2^n}, d_0 + \frac{\pi}{2^n}) : f(t) > c \}) \\
&= \quad 2^n m(E \cap (d_0 - \frac{\pi}{2^n}, d_0 + \frac{\pi}{2^n})) \\
&> \quad 1.7\pi,
\end{aligned}$$

a contradiction. $\square$

Theorem 3.11. *Let ψ be an orthogonal wavelet whose Fourier transformation is non-zero a.e. Then*

$$\{D, T, P_\psi\}' = \mathbb{C}I.$$

Proof. Let $W \in \{D, T, P_\psi\}'$ be arbitrary. By Theorem 3.5, $\widehat{W} = M_h$ for some

2-dilation-periodic function h. Since W commutes with P_ψ we have $W\psi \in P_\psi \mathcal{H}$, so $W\psi = \sum_{n \in \mathbb{Z}} \alpha_n T^n \psi$ for some $(\alpha_n) \in l_2(\mathbb{Z})$. Thus

$$h\widehat{\psi} = \sum_{n \in \mathbb{Z}} \alpha_n \widehat{T}^n \widehat{\psi} = \sum_{n \in \mathbb{Z}} \alpha_n e^{-ins} \widehat{\psi} \quad \text{a.e.}$$

Let f be the 2π-periodic function given by the sum $\sum_{n \in \mathbb{Z}} \alpha_n e^{-ins}$, where convergence is in $L^2[0, 2\pi]$ with periodic extension to $\mathbb{R}$. It follows that

$$h(s)\widehat{\psi}(s) = f(s)\widehat{\psi}(s) \text{ a.e.,}$$

and since $\widehat{\psi}(s) \neq 0$ a.e. we must have $h(s) = f(s)$ a.e. Thus h is 2π-periodic as well as 2- dilation-periodic, and hence constant (a.e.) by Lemma 3.10. So $W \in \mathbb{C}I$. $\square$

If ψ has compact support, such as the Haar wavelet and Daubechies wavelet, then $\widehat{\psi}$ has an analytic extension to the complex plane, so the hypothesis of Theorem 3.11 are satisfied.

Corollary 3.12. *If $\widehat{\psi}$ is non-zero (a.e.), let $V \in \mathbb{U}(\{D, T\}') \setminus \mathbb{C}I$. Then $V\psi \notin E_\psi$.*

Proof. Suppose that $V \in \{D, T\}'$ and that $V\psi \in E_\psi$. Then Theorem 3.9 implies that $P_\psi V = V P_\psi$, and thus $V \in \{D, T, P_\psi\}'$. So the above theorem implies that $V \in \mathbb{C}I$. $\square$

The method of multiresolution analysis is important in wavelet theory. For references, see, for instance [**19, 21**] or [**8**].

Corollary 3.13. Let ψ be an orthogonal wavelet with compact support. Suppose that ψ is generated by a multiresolution analysis. Let

$$V \in \mathbb{U}(\{D, T\}') \setminus \mathbb{C}I$$

and $\eta = V\psi$. Then η is *not* generated by the same multiresolution analysis.

Proof. Wavelets ψ_1, ψ_2 with the same multiresolution analysis are related in that the Fourier transform of one can be obtained from the Fourier transform of the other by multiplication by a 2π- translation-invariant unimodular function. (c.f. [**8**]. See also Notes 3.8 (ii).) Hence $E_{\psi_1} = E_{\psi_2}$. $\square$

The algebra $\{D, T, P_\psi\}'$ can be regarded as an operator algebraic unitary invariant for wavelets. If ψ does not satisfy the hypotheses of Theorem 3.11, then $\{D, T, P_\psi\}'$ need not be trivial. For instance, if E is a *wavelet set* and E is the corresponding s-elementary wavelet (see Chapter 4) then $\widehat{P_\psi} = M_{\chi_E}$, a multiplication operator. So since $\{\widehat{D}, \widehat{T}\}'$ consists of multiplication operators, in this case we have

$$\{D, T, P_\psi\}' = \{D, T\}'.$$

In other cases $\{D, T, P_\psi\}'$ may lie strictly between $\mathbb{C}I$ and $\{D, T\}'$. (See Corollary 3.18, Corollary 5.12 and Example 5.13.)

The methods of chapter 2 yield a generalization of Theorem 3.11.

Theorem 3.14. *With the hypotheses of Theorem 2.16, if $\psi \in \mathcal{W}(\mathcal{U})$ then*

$$\{\mathcal{U}, P_\psi\}' = \{A \in \mathcal{U}' : \exists\ B \in w^*(\mathcal{U}_0) \text{ with } (A - B)\psi = 0\}.$$

Proof. Let $A \in \mathcal{U}'$. If there exists $B \in w^*(\mathcal{U}_0)$ with $(A - B)\psi = 0$, then $A\psi = B\psi = \kappa_\psi(B)\psi$. So since A, $\kappa_\psi(B) \in \mathcal{C}_\psi(\mathcal{U})$ and ψ is separating for $\mathcal{C}_\psi(\mathcal{U})$,

$A = \kappa_\psi(B)$. Then $A^* = \kappa_\psi(B^*)$, so $(A^* - B^*)\psi = 0$. So $A\psi \in P_\psi \mathcal{H}$ and $A^*\psi \in P_\psi \mathcal{H}$. It follows that A reduces $P_\psi \mathcal{H}$, so commutes with P_ψ, as required. Conversely, if $A \in \{\mathcal{U}, P_\psi\}'$, then $A \in \mathcal{C}_\psi(\mathcal{U})$ and $AP_\psi = P_\psi A$. Thus by Theorem 2.16 there exists $B \in \mathrm{w}^*(\mathcal{U}_0)$ with $A\psi = B\psi$. $\square$

Corollary 3.15. Let ψ be an arbitrary orthogonal wavelet. Then $A \in \{D, T, P_\psi\}'$ if and only if $\widehat{A} = M_f$, where f is a bounded 2-dilation-periodic function which coincides on the support of $\widehat{\psi}$ with a 2π-translation-periodic function.

Proof. By Theorem 3.14, $\{A \in D, T, P_\psi\}'$ iff there exists $B \in \mathrm{w}^*(T)$ with $(A - B)\psi = 0$. But $A \in \{D, T\}'$ implies $\widehat{A} = M_f$ for some 2-dilation-periodic L^∞-function, and $B \in \mathrm{w}^*(T)$ implies $\widehat{B} = M_g$ for some 2π-translation-periodic function, as required. (In case $\mathrm{supp}(\widehat{\psi}) = \mathbb{R}$ this implies $f = g$, which then must be constant by Lemma 3.10, recovering Theorem 3.11.) $\square$

To establish density of $\mathrm{span}(\mathcal{W}(D, T))$ we require a lemma.

Proposition 3.16 . *The von Neumann algebra generated by*

$\{M_f : f \in L^\infty(\mathbb{R}), f \text{ is } 2\pi\text{-periodic}\}$ and $\{M_g : g \in L^\infty(\mathbb{R}), g \text{ is } 2\text{-dilation-periodic}\}$

is

$$\mathcal{D} := \{M_h : h \in L^\infty(\mathbb{R})\}.$$

Proof. It will be convenient to work with 1-periodic, rather then 2π-periodic, functions. If we let $W = D_{2\pi}$, then $WD = DW$, and $WM_{e^{ins}}W^* = M_{e^{2\pi ins}}$, so

$$W\{M_f : f \text{ is } 2\pi\text{-periodic }\}W^* = \{M_f : f \text{ is } 1\text{-periodic}\}.$$

So the problems are equivalent.

Let $\mathcal{A}$ be the von Neumann algebra generated by the 1-periodic and 2-dilation periodic multiplication operators. First, suppose $k \in \mathbb{Z}, k \geq 1$, and let $E \subset [k, k+1)$ be an arbitrary measurable subset. Let n be the (nonnegative) integer such that $[2^n, 2^{n+1})$ contains $[k, k+1)$, and define a 2-dilation periodic function $g(s)$ on $\mathbb{R}$ by $g(s) = s - k$ for $s \in E, g(s) = 0$ for $s \in [2^n, 2^{n+1}) \setminus E$, $g(s) = g(2^{-l}s)$ for $s \in 2^l[2^n, 2^{n+1}), l \in \mathbb{Z}$, and $g(s) = 0$ for $s \leq 0$.

Now define $h(s)$ on $[0, 1)$ by $h(s) = s$, and extend 1-periodically to $\mathbb{R}$. Let $r(s) = h(s) - g(s)$.

On each interval $[l, l+1)$ for $l \in \mathbb{Z}, l \geq 1$, the graph of $g(s)$ is 0 or a portion of a straight line. The only such interval on which the graph of $g(s)$ is a portion of a straight line of *slope* 1 is $[k, k+1)$. On each such interval $[l, l+1)$, the graph of $h(s)$ is a straight line of slope 1. So on each interval $[l, l+1), l \geq 1, l \neq k$, the graphs of g and h intersect in at most one point. On $[k, k+1)$, the graphs of g and h coincide on E and differ on $[k, k+1) \setminus E$. On $[0, 1)$, the graph of $g(s)$ is piecewise defined by straight lines, none of slope 1, and the graph of $h(s)$ is a straight line of slope 1, so the graphs of g and h intersect at countably many points on $[0, 1)$.

Since $h(s)$ is nonzero except on the integers, the above paragraph shows that $r(s) = 0$ on E and at countably many points not in E. Thus the kernel projection of M_r is M_{χ_E}. So $M_{\chi_E} \in \mathcal{A}$ since $M_r \in \mathcal{A}$.

A construction analogous to the above shows that if $k \in \mathbb{Z}$ with $k \leq -2$ and $E \subseteq [k, k+1)$ then $M_{\chi_E} \in \mathcal{A}$.

Next consider measurable subsets of $[-1, 1)$. If E is a subset of $[2^{-(k+1)}, 2^{-k})$ for some $k \geq 0$, then an argument similar to the above shows that $M_{\chi_E} \in \mathcal{A}$. The same is true for a subset E of $[-2^{-k}, -2^{-(k+1)})$.

Every measurable subset of $\mathbb{R}$ is a union of countably many sets of the form E considered above. So $M_{\chi_F} \in \mathcal{A}$ for every measurable set $F \subseteq \mathbb{R}$, and hence $\mathcal{A} = \mathcal{D}$. $\square$

Corollary 3.17 . *The linear span of* $\mathcal{W}(D, T)$ *is dense in* $L^2(\mathbb{R})$.

Proof. Let ψ be any wavelet with the property that $\widehat{\psi}(s)$ is nonzero a.e. Let $\mathcal{E}$ be the linear span of the operators M_h with h of the form $h = fg$ where f is unimodular and 2π-periodic, and g is unimodular and 2-dilation periodic. Then Proposition 3.16 implies that $\mathcal{D}$ is the strong operator topology closure of $\mathcal{E}$. For any such h, $\mathcal{F}^{-1}(h\widehat{\psi})$ is a wavelet. [See Remark 3.6 and 3.7.] Thus $\mathcal{F}^{-1}(\mathcal{E}\widehat{\psi})$ is a linear span of wavelets. Since $\widehat{\psi}$ is nonzero a.e. it is cyclic for $\mathcal{D}$. Thus $[\mathcal{E}\widehat{\psi}] = \mathcal{H}$, and so $[\mathcal{F}^{-1}(\mathcal{E}\widehat{\psi})] = \mathcal{H}$, as required. $\square$

We conclude with a problem.

Problem D. *Suppose* ψ *and* η *are orthogonal wavelets such that* $|\widehat{\psi}(s)| = |\widehat{\eta}(s)|$ *a.e. Do there exist a* 2π-*translation-periodic unimodular function* $h(s)$ *and a* 2-*dilation-periodic unimodular function* $g(s)$ *such that*

$$\widehat{\eta}(s) = h(s)g(s)\widehat{\psi}(s)?$$

CHAPTER 4

Wavelet Sets

We say that measurable sets E, F are *translation congruent modulo 2π* if there is a measurable bijection $\phi : E \to F$ such that $\phi(s) - s$ is an integral multiple of 2π for each $s \in E$; or equivalently, if there is a measurable partition $\{E_n : n \in \mathbb{Z}\}$ of E such that $\{E_n + 2n\pi : n \in \mathbb{Z}\}$ is a measurable partition of F. Analogously, define measurable sets G and H to be *dilation congruent modulo 2* if there is a measurable bijection $\tau : G \to H$ such that for each $s \in G$ there is an integer n, depending on s, such that $\tau(s) = 2^n s$; or equivalently, if there is a measurable partition $\{G_n\}_{-\infty}^{\infty}$ of G such that $\{2^n G_n\}_{-\infty}^{\infty}$ is a measurable partition of H. (Translation and dilation congruency modulo other positive numbers of course make sense as well.)

Lemma 4.1. *Let $f \in L^2(\mathbb{R})$, and let $E = \mathrm{supp}(f)$. Then f has the property that $\{e^{ins} f : n \in \mathbb{Z}\}$ is an orthonormal basis for $L^2(E)$ if and only if*

1. *E is congruent to $[0, 2\pi)$ modulo 2π, and*
2. *$|f(s)| = \frac{1}{\sqrt{2\pi}}$ a.e. on E.*

Proof. If $E \sim [0, 2\pi)$ mod 2π, then clearly

$$\{\frac{1}{\sqrt{2\pi}} e^{ins}|_E : n \in \mathbb{Z}\}$$

is an o.n. basis for $L^2(E)$. If this is multiplied by a unimodular function it remains a basis. This completes the "only if" part.

For the converse, suppose $\{e^{ins} f\}$ is an o.n. basis. Firstly, assume by way of contradiction that E is *not* translation congruent to a *subset* of $[0, 2\pi)$ modulo 2π. Then there is a subset $F \subset E$ of finite positive measure such that $F + 2\pi k \subset E$ for some nonzero integer k. Replacing F with a subset if necessary, we can assume $F \cap (F + 2\pi k) = \emptyset$, and also that $f|_F$ and $f|_{F+2\pi k}$ are bounded. Let $G = F \cup (F + 2\pi k)$. Then $\chi_G f^2 \in L^2(E)$, so we may expand

$$\chi_G f^2 = \sum \alpha_n e^{ins} f,$$

with $(\alpha_n) \in l^2(\mathbb{Z})$. Let g be the 2π-periodic function on $\mathbb{R}$ defined by $\sum \alpha_n e^{ins}$ on $[0, 2\pi)$ and extended periodically. Then

$$\chi_G f^2 = fg \text{ (a.e.) on } \mathbb{R}.$$

Since f is nonzero on E, this implies $\chi_G f = \chi_E g$ (a.e.). So since g is 2π-periodic, $f(s + 2\pi k) = f(s)$ (a.e.) on F. Let $f_n = e^{ins} f$. Then also $f_n(s + 2\pi k) = f_n(s)$ (a.e.) on F. Hence each f_n is orthogonal to the nonzero function

$$h = \chi_F - \chi_{F+2\pi k}.$$

This contradicts the fact that $\{f_n : n \in \mathbb{Z}\}$ is an orthonormal basis for $L^2(E)$. We have proven that $E \sim E'$ mod 2π, where E' is a subset of $[0, 2\pi)$.

Let $\phi : E' \to E$ be the bijection establishing congruency. Then

$$f_n \circ \phi = e^{ins}(f \circ \phi)$$

is an orthonormal basis for $L^2(E')$. Regarding these as functions in $L^2[0, 2\pi]$ with support E', this implies

$$\int_0^{2\pi} |f \circ \phi|^2 e^{ils} ds = 0, l \in \mathbb{Z}, l \neq 0.$$

Hence $|f \circ \phi|$ must be constant on $[0, 2\pi)$. This implies $E' = [0, 2\pi)$ modulo a null set, so $E \sim [0, 2\pi)$ mod 2π, and also that $|f|$ must be constant on E. Since $|f|$ is a unit vector, this constant is $\frac{1}{\sqrt{2\pi}}$, as required. $\square$

Let us define a measurable subset $E \subset \mathbb{R}$ to be a *wavelet set* if $\frac{1}{\sqrt{2\pi}}\chi_E$ is the Fourier transform of a wavelet. We will call such wavelets *s-elementary wavelets*. (The prefix "s" is for "set".) We use the notation

$$\widehat{\psi}_E := \frac{1}{\sqrt{2\pi}}\chi_E.$$

The classic example is given by the Littlewood-Paley orthonormal basis $\{2^{\frac{n}{2}}\widehat{\psi}(2^n - l) : n, l \in \mathbb{Z}\}$ with $\widehat{\psi} = \widehat{\psi}_E$ for $E = [-2\pi, -\pi) \cup [\pi, 2\pi)$. (c.f. [8], p.115). This set E is translation-congruent modulo 2π to $[0, 2\pi)$, since $[-2\pi, -\pi) + 2\pi = [0, \pi)$. So since $\{\frac{1}{\sqrt{2\pi}}e^{ils} : l \in \mathbb{Z}\}$ is an orthonormal basis for $L^2([0, 2\pi))$, and since $\widehat{T} = M_{e^{is}}$ on $\mathbb{R}$, it follows that $\{\widehat{T}^l\widehat{\psi} : l \in \mathbb{Z}\}$ is an orthonormal basis for $L^2(E)$. If f is a function with support in E, then $D^n f$ has support in $2^{-n}E$. Since $\widehat{D} = D^{-1}$, it follows that for each n, $\{\widehat{D}^n\widehat{T}^l\widehat{\psi} : l \in \mathbb{Z}\}$ is an orthonormal basis for $L^2(2^n E)$. Since the sets $\{2^n E : n \in \mathbb{Z}\}$ are disjoint and have union $\mathbb{R} \setminus \{0\}$, it follows that $\{\widehat{D}^n\widehat{T}^l\widehat{\psi} : n, l \in \mathbb{Z}\}$ is an orthonormal basis for $L^2(\mathbb{R})$. That is, ψ is a wavelet. From this argument, it is clear that if E is *any* measurable set which is both translation congruent to $[0, 2\pi)$ modulo 2π *and* has the property that $\{2^n E : n \in \mathbb{Z}\}$ is a partition of $\mathbb{R}$ (modulo a null set) then E is a wavelet set. We will prove the converse. But first, it is convenient to describe how these properties are essentially related.

We say that a measurable subset $G \subset \mathbb{R}$ is a *2-dilation generator of a partition* of $\mathbb{R}$ if the sets

$$2^n G := \{2^n s : s \in G\}, n \in \mathbb{Z}$$

are disjoint and $\mathbb{R} \setminus \cup_n 2^n G$ is a null set. Also, we say that $E \subseteq \mathbb{R}$ is a *2π-translation generator of a partition* of $\mathbb{R}$ if the sets

$$E + 2n\pi := \{s + 2n\pi : s \in E\}, n \in \mathbb{Z},$$

are disjoint and $\mathbb{R} \setminus \cup_n (E + 2n\pi)$ is a null set.

Lemma 4.2. *A measurable set $E \subseteq \mathbb{R}$ is a 2π-translation generator of a partition of $\mathbb{R}$ if and only if, modulo a null set, E is translation congruent to $[0, 2\pi)$ modulo 2π. Also, a measurable set $G \subseteq \mathbb{R}$ is a 2-dilation generator of a partition of $\mathbb{R}$ if and only if, modulo a null set, G is dilation congruent modulo 2 to the set $[-2\pi, -\pi) \cup [\pi, 2\pi)$.*

Proof.   Suppose E is a translation generator. For each $s \in E$ let $n(s)$ be the unique integer with $s - 2\pi n(s) \in [0, 2\pi)$, and define

$$\phi(s) = s - 2\pi n(s), s \in E.$$

Disjointness of the sets $(E + 2\pi n)$ implies that ϕ is 1-1 on E, and the covering of $\mathbb{R}$ implies that ϕ is onto $[0, 2\pi)$. (Modulo null sets in both cases, of course.) Conversely, if $E \sim [0, 2\pi)$ modulo 2π it is obvious that E is a generator.

Similarly, if G is a 2-dilation generator, then for each $s \in G, s \neq 0$, let $n(s)$ be the unique integer such that

$$2^{n(s)} s \in [-2\pi, -\pi) \cup [\pi, 2\pi),$$

and define

$$\tau(s) = 2^{n(s)} s, s \in G.$$

Disjointness of $\{2^l G : l \in \mathbb{Z}\}$ implies that τ is 1-1, and covering implies τ is onto $[-2\pi, -\pi) \cup [\pi, 2\pi)$. The converse is obvious. $\square$

Lemma 4.3.   *Let $E \subseteq \mathbb{R}$ be a measurable set. Then E is a wavelet set if and only if E is both a 2-dilation generator of a partition (modulo null sets) of $\mathbb{R}$ and a 2π-translation generator of a partition (modulo null sets) of $\mathbb{R}$. Equivalently, E is a wavelet set if and only if E is both translation congruent to $[0, 2\pi)$ modulo 2π and dilation congruent to $[-2\pi, -\pi) \cup [\pi, 2\pi)$ modulo 2.*

Proof.   The "if" part is obvious from Lemma 4.2. For the converse, let E be a wavelet set. Then

$$\widehat{D}^n \widehat{\psi}_E = \frac{1}{\sqrt{2\pi}} \chi_{2^{-n} E}.$$

Since these are orthogonal the sets $\{2^n E : n \in \mathbb{Z}\}$ must be disjoint. It follows that

$$\{\widehat{T}^l \widehat{\psi}_E : l \in \mathbb{Z}\} = \{\frac{1}{\sqrt{2\pi}} e^{ils} \chi_E : l \in \mathbb{Z}\}$$

is an orthonormal basis for $L^2(E)$, and hence by Lemma 4.1 that E is translation-congruent to $[0, 2\pi)$ modulo 2π. Since $\{\widehat{D}^n \widehat{T}^l \widehat{\psi}_E : n, l \in \mathbb{Z}\}$ is a basis the union of the supports is full, and hence $\{2^n E : n \in \mathbb{Z}\}$ must be a partition of $\mathbb{R}$. $\square$

Remark 4.4.   If E is a wavelet set, and if $f(s)$ is any function with support E which has constant modulus $\frac{1}{\sqrt{2\pi}}$ on E, then $\mathcal{F}^{-1}(f)$ is a wavelet. Indeed, by Lemma 4.1 $\{\widehat{T}^l f :\in \mathbb{Z}\}$ is an orthonormal basis for $L^2(E)$, and since the sets $2^n E$ partition $\mathbb{R}$, it follows that $\{\widehat{D}^n \widehat{T}^l f : n, l \in \mathbb{Z}\}$ must be an orthonormal basis for $L^2(\mathbb{R})$, as required.

Example 4.5.   It is usually easy to determine, using the dilation-translation criteria, whether a given finite union of intervals is a wavelet set. On the other hand wavelet sets, suitable for testing hypotheses, can be quite difficult to construct. We present some of these, both for usefulness in the sequel, and for perspective here. Items (ii) $\to$ (ix) appear to be new.

(i) An example due to Journe (c.f. [**8**], p.136) of a wavelet which admits no multiresolution analysis is the s-elementary wavelet with wavelet set

$$[-\frac{32\pi}{7}, -4\pi) \cup [-\pi, -\frac{4\pi}{7}) \cup [\frac{4\pi}{7}, \pi) \cup [4\pi, \frac{32\pi}{7}).$$

To see that this satisfies the criteria, label these intervals, in order, as J_1, J_2, J_3, J_4 and write $J = \cup J_i$. Then

$$J_1 \cup 4J_2 \cup 4J_3 \cup J_4 = [-\frac{32\pi}{7}, -\frac{16\pi}{7}) \cup [\frac{16\pi}{7}, \frac{32\pi}{7}).$$

This has the form $[-2a, -a) \cup [b, 2b)$ so is a 2-dilation generator of a partition of $\mathbb{R} \setminus \{0\}$. Then also observe that

$$\{J_1 + 6\pi, \ J_2 + 2\pi, \ J_3, \ J_4 - 4\pi\}$$

is a partition of $[0, 2\pi)$.

(ii) The Littlewood-Paley set can be generalized. For any $-\pi < \alpha < \pi$, the set

$$E_\alpha = [-2\pi + 2\alpha, -\pi + \alpha) \cup [\pi + \alpha, 2\pi + 2\alpha)$$

is a wavelet set. Indeed, it is clearly a 2-dilation generator of a partition of $\mathbb{R} \setminus \{0\}$, and to see that it satisfies the translation congruency criterion for $-\pi < \alpha \leq 0$ (the case $0 < \alpha < \pi$ is analogous) just observe that

$$\{[-2\pi + 2\alpha, -2\pi) + 4\pi, \ [-2\pi, -\pi + \alpha) + 2\pi, \ [\pi + \alpha, 2\pi + 2\alpha)\}$$

is a partition of $[0, 2\pi)$. It is clear that ψ_{E_α} is then a continuous (in $L^2(\mathbb{R})$-norm) path of s-elementary wavelets. Note that

$$\lim_{\alpha \to \pi} \widehat{\psi}_{E_\alpha} = \frac{1}{\sqrt{2\pi}} \chi_{[2\pi, 4\pi)}.$$

This is *not* the Fourier transform of a wavelet because the set $[2\pi, 4\pi)$ is not a 2-dilation generator of a partition of $\mathbb{R} \setminus \{0\}$. So

$$\lim_{\alpha \to \pi} \psi_{E_\alpha}$$

is not an orthogonal wavelet. (It is what is known as a Hardy wavelet because it generates an orthonormal basis for $H^2(\mathbb{R})$ under dilation and translation.) This example demonstrates that $\mathcal{W}(D, T)$ is *not* closed in $L^2(\mathbb{R})$.

(iii) Journe's example above can be extended to a path. For $-\frac{\pi}{7} \leq \beta \leq \frac{\pi}{7}$ the set

$$J_\beta = [-\frac{32\pi}{7}, -4\pi + 4\beta) \cup [-\pi + \beta, -\frac{4\pi}{7}) \cup [\frac{4\pi}{7}, \pi + \beta) \cup [4\pi + 4\beta, 4\pi + \frac{4\pi}{7})$$

is a wavelet set. The same argument in (i) establishes dilation congruency. For translation, the argument in (i) shows congruency to $[4\beta, 2\pi + 4\beta)$ which is in turn congruent to $[0, 2\pi)$ as required. Observe that here, as opposed to in (ii) above, the limit of ψ_{J_β} as β approaches the boundary point $\frac{\pi}{7}$ *is* a wavelet. Its wavelet set is a union of 3 disjoint intervals.

(iv) While the Littlewood-Paley and the Journe wavelet sets are *symmetric* by reflection through the origin (modulo the boundary, which is a null set), the paths in (ii) and (iii) consist of non-symmetric sets (except at 0). It is noteworthy that paths of *symmetric* wavelet sets also exist: For example, consider for $0 \leq \alpha \leq \frac{\pi}{3}$,

$$F_\alpha = [-\frac{8\pi}{3} + 2\alpha, -2\pi) \cup [-\frac{4\pi}{3} - 2\alpha, -\frac{4\pi}{3} + \alpha) \cup [-\pi, -\frac{2\pi}{3} - \alpha)$$

$$\cup [\frac{2\pi}{3} + \alpha, \pi) \cup [\frac{4\pi}{3} - \alpha, \frac{4\pi}{3} + 2\alpha) \cup [2\pi, \frac{8\pi}{3} - 2\alpha).$$

We leave to the reader the (easy) verification that F_α satisfies the dilation and translation congruency criteria so is a wavelet set. Note that $F_{\frac{\pi}{3}}$ is the Littlewood-Paley set. We have

$$F_0 = [-\frac{8\pi}{3}, -2\pi) \cup [-\pi, -\frac{2\pi}{3}) \cup [\frac{2\pi}{3}, \pi) \cup [2\pi, \frac{8\pi}{3}).$$

(v) The wavelet set

$$[-\frac{\pi}{2}, -\frac{\pi}{4}) \cup [\pi, \frac{5\pi}{4}) \cup [\frac{7\pi}{4}, 2\pi) \cup [\frac{5\pi}{2}, 3\pi) \cup [\frac{13\pi}{4}, \frac{7\pi}{2}) \cup [6\pi, \frac{13\pi}{2})$$

is the union of 6 disjoint intervals, all but one of which are positive. This illustrates that wavelet sets can be very asymmetric in structure.

(vi) Let $0 < \alpha < \beta < \gamma < \delta < \cdots < \frac{\pi}{2}$. The sets of item (ii) admit further "splitting" into multiparameter families of wavelet sets:

$$\begin{aligned} E_{\alpha\beta} \quad &= [-2\pi, -2\pi + 2\alpha) \cup [-2\pi + 2\beta, -\pi) \cup [-\pi + \alpha, -\pi + \beta) \\ &\cup [\pi, \pi + \alpha) \cup [\pi + \beta, 2\pi) \cup [2\pi + 2\alpha, 2\pi + 2\beta) \end{aligned}$$

$$\begin{aligned} E_{\alpha\beta\gamma} \quad &= [-2\pi + 2\alpha, -2\pi + 2\beta) \cup [-2\pi + 2\gamma, -\pi + \alpha) \cup [-\pi + \beta, -\pi + \gamma) \\ &\cup [\pi + \alpha, \pi + \beta) \cup [\pi + \gamma, 2\pi + 2\alpha) \cup [2\pi + 2\beta, 2\pi + 2\gamma) \end{aligned}$$

$$\begin{aligned} E_{\alpha\beta\gamma\delta} \quad &= [-2\pi, -2\pi + 2\alpha) \cup [-2\pi + 2\beta, -2\pi + 2\gamma) \cup [-2\pi + 2\delta, -\pi) \\ &\cup [-\pi + \alpha, -\pi + \beta) \cup [-\pi + \gamma, -\pi + \delta) \cup [\pi, \pi + \alpha) \\ &\cup [\pi + \beta, \pi + \gamma) \cup [\pi + \delta, 2\pi) \cup [2\pi + 2\alpha, 2\pi + 2\beta) \\ &\cup [2\pi + 2\gamma, 2\pi + 2\delta). \end{aligned}$$

This process can be continued. It is perhaps curious that $E_{\alpha\beta}$ and $E_{\alpha\beta\gamma}$ have 6 disjoint intervals, yet $E_{\alpha\beta\gamma\delta}$ has 10. It will be shown (see example A.4) that these arise naturally from operator-interpolation starting with the Littlewood-Paley set E_0 and the family $\{E_\alpha : 0 < \alpha < \frac{\pi}{2}\}$.

(vii) Another curious easily-checked family of wavelet sets is

$$\begin{aligned} G_\alpha \quad &= \quad [-\frac{8\pi}{3}, -\frac{8\pi}{3} + 2\alpha) \cup [-\frac{4\pi}{3} + \alpha, -\frac{2\pi}{3}) \\ &\cup [\frac{2\pi}{3}, \frac{2\pi}{3} + \alpha) \cup [\frac{4\pi}{3} + 2\alpha, \frac{8\pi}{3}) \end{aligned}$$

for $0 \le \alpha \le \frac{\pi}{3}$. Note that these are simple perturbations of the set E which is used in the analysis of Meyer's family ψ_{Me} in Proposition 5.5. (We thank Eugen Ionascu for this example.)

(viii) Let $A \subseteq [\pi, \frac{3\pi}{2})$ be an arbitrary measurable subset. Then there is a wavelet set W, such that $W \cap [\pi, \frac{3\pi}{2}) = A$. For the construction, let

$$B = [2\pi, 3\pi) \setminus 2A,$$

$$C = [-\pi, -\frac{\pi}{2}) \setminus (A - 2\pi)$$

and $\quad D = 2A - 4\pi.$

Let

$$W = [\frac{3\pi}{2}, 2\pi) \cup A \cup B \cup C \cup D.$$

We have $W \cap [\pi, \frac{3\pi}{2}) = A$. Observe that the sets $[\frac{3\pi}{2}, 2\pi)$, A, B, C, D, are disjoint. Also observe that the sets

$$[\frac{3\pi}{2}, 2\pi), \ A, \ \frac{1}{2}B, \ 2C, \ D,$$

are disjoint and have union $[-2\pi, -\pi) \cup [\pi, 2\pi)$. In addition, observe that the sets

$$[\frac{3\pi}{2}, 2\pi), \ A, \ B - 2\pi, \ C + 2\pi, \ D + 2\pi,$$

are disjoint and have union $[0, 2\pi)$. Hence W is a wavelet set.

(ix) Let $A \subseteq (\frac{8\pi}{3}, 3\pi)$ be an arbitrary measurable subset. Then there is a *symmetric* (by reflection through the origin) wavelet set W, such that $W \cap (\frac{8\pi}{3}, 3\pi) = A$. For the construction, let

$$B = -\frac{1}{2}A + 2\pi \ \text{ and } \ C = [\pi, 2\pi) \setminus (2B \cup \frac{1}{2}A).$$

We claim that the symmetric set

$$W = -(A \cup B \cup C) \cup (A \cup B \cup C)$$

satisfies our requirements. Observe that the sets A, B, C are disjoint and contained in $(0, \infty)$. Then observe that the sets $\frac{1}{2}A$, $2B$, C are disjoint and have union $[\pi, 2\pi)$, so W is 2-dilation congruent to $[-2\pi, -\pi) \cup [\pi, 2\pi)$ modulo a null set. Then note that

$$\frac{1}{2}A = -B + 2\pi \ \text{ and } \ 2B = -A + 4\pi.$$

So the sets $-A + 4\pi$, $-B + 2\pi$, C are disjoint and have union $[\pi, 2\pi)$, and the sets $A - 4\pi$, $B - 2\pi$, $-C$ are disjoint and have union $[-2\pi, -\pi)$. So W is 2π-translation congruent to $[-2\pi, -\pi) \cup [\pi, 2\pi)$, and hence to $[0, 2\pi)$. This shows that W is a wavelet set. By the construction we have $W \cap (\frac{8\pi}{3}, 3\pi) = A$.

(x) Wavelet sets for arbitrary (not necessarily integral) dilation factors other then 2 exist. For instance, if $d \geq 2$ is arbitrary, let

$$A = [-\frac{2d\pi}{d+1}, -\frac{2\pi}{d+1}),$$

$$B = [\frac{2\pi}{d^2-1}, \frac{2\pi}{d+1}),$$

$$C = [\frac{2d\pi}{d+1}, \frac{2d^2\pi}{d^2-1})$$

and let $G = A \cup B \cup C$. Then G is a d-wavelet set. To see this, note that $\{A + 2\pi, B, C\}$ is a partition of an interval of length 2π. So G is 2π-translation-congruent to $[0, 2\pi)$. Also, $\{A, B, d^{-1}C\}$ is a partition of the set $[-d\alpha, -\alpha) \cup [\beta, d\beta)$ for $\alpha = \frac{2\pi}{d^2-1}$, and $\beta = \frac{2\pi}{d^2-1}$, so from this form it follows that $\{d^n G : n \in \mathbb{Z}\}$ is a partition of $\mathbb{R} \setminus \{0\}$. Hence if $\psi := \mathcal{F}^{-1}(\frac{1}{\sqrt{2\pi}}\chi_G)$, it follows that $\{d^{\frac{n}{2}}\psi(d^n t - l) : n, l \in \mathbb{Z}\}$ is an orthonormal basis for $L^2(\mathbb{R})$, as required. For dilation factors $1 < d < 2$, a similar type of construction yields a d-wavelet set. (We thank Puhong You for this example.)

(xi) There exist unbounded wavelet sets. Let $\{A_n : n = 0, 1, 2, \cdots\}$ be a measurable partition of $[\pi, 2\pi)$. Then the sets $\{A_n\}$ are disjoint, and for $n \geq 1$ we have $2^{-n} A_n \subseteq [0, \pi)$. Let $B_1 = \cup_{n=1}^{\infty} 2^{-n} A_n$. Then let

$$
\begin{aligned}
B &= \cup_{n=0}^{\infty} 2^{-n} A_n \\
C &= [-2\pi, -\pi) \backslash (B_1 - 2\pi) \\
D &= \cup_{n=1}^{\infty} 2^n (2^{-n} A_n - 2\pi).
\end{aligned}
$$

Note that the set D is unbounded. We leave to the reader the verification that

$$
W = B \cup C \cup D
$$

satisfies the dilation-translation congruency criteria so is a wavelet set. (We thank Eugen Ionascu for this example.)

The following gives an operator algebraic characterization of s-elementary wavelets. (See also Remark 4.4 above.)

Theorem 4.6. *Let* $\psi \in \mathcal{W}(\mathcal{U}_{D,T})$. *Let* $V_\psi = \kappa_\psi(T)$ *be the unique unitary operator in* $\mathcal{C}_\psi(\mathcal{U}_{D,T})$ *with* $V_\psi \psi = T\psi$. *Then* $TV_\psi = V_\psi T$ *if and only if*

$$
|\widehat{\psi}| = \frac{1}{\sqrt{2\pi}} \chi_E
$$

for some wavelet set E.

Proof. Suppose E is a wavelet set. Define a unimodular function $h(s)$ on $\mathbb{R}$ by setting $h(s) = e^{-is}$ for $s \in E$ and extending 2-dilation periodically by

$$
h(s) = h(2^{-n} s), \quad s \in 2^n E, \quad n \in \mathbb{Z},
$$

and $h(0) = 1$. Then $M_h \in \{D, T\}' \subset \mathcal{C}_\psi(D, T)$, and $M_h \widehat{\psi} = \widehat{T}\widehat{\psi}$, so by uniqueness we must have $M_h = \widehat{V}_\psi$. Since $\widehat{T} = M_{e^{-is}}$, the operators $\widehat{T}$ and M_h commute, completing the "if" part.

For the converse, assume that $TV_\psi = V_\psi T$. By Lemma 3.1, V_ψ commutes with D also. So $V_\psi \in \{D, T\}'$. Thus $\widehat{V}_\psi$ is a multiplication operator, by Theorem 3.5. Let $\beta \in \mathbb{R}$. It follows that V_ψ commutes with T_β. Let $n, l \in \mathbb{Z}$. We compute

$$
\begin{aligned}
\langle V_\psi T_\beta \psi, D^n T^l \psi \rangle &= \langle T_\beta \psi, V_\psi^* D^n T^l \psi \rangle \\
&= \langle T_\beta \psi, D^n T^l V_\psi^* \psi \rangle \\
&= \langle T_\beta \psi, D^n T^{l-1} \psi \rangle.
\end{aligned}
$$

By Lemma 3.2, $D^n T^{-1} = T_{-2^{-n}} D^n$, so $D^n T^{l-1} = T_{-2^{-n}} D^n T^l$. Thus

$$
\langle V_\psi T_\beta \psi, D^n T^l \psi \rangle = \langle T_\beta \psi, T_{-2^{-n}} D^n T^l \psi \rangle.
$$

Also,

$$
\begin{aligned}
\langle T_\beta V_\psi \psi, D^n T^l \psi \rangle &= \langle T_\beta T\psi, D^n T^l \psi \rangle \\
&= \langle TT_\beta \psi, D^n T^l \psi \rangle \\
&= \langle T_\beta \psi, T^{-1} D^n T^l \psi \rangle.
\end{aligned}
$$

It follows that

$$
\langle T_\beta \psi, (T_{-2^{-n}} - T^{-1}) D^n T^l \psi \rangle = 0
$$

for all $n, l \in \mathbb{Z}$ and $\beta \in \mathbb{R}$. Now take the Fourier transform of this equation. We have

$$\langle \widehat{T}_\beta \widehat{\psi}, (\widehat{T}_{-2^{-n}} - \widehat{T}^{-1}) \widehat{D}^n \widehat{T}^l \widehat{\psi} \rangle = 0, \ \beta \in \mathbb{R},$$

for each fixed $n, l \in \mathbb{Z}$. Since $\{\widehat{T}_\beta : \beta \in \mathbb{R}\}$ generates

$$\mathcal{D}(\mathbb{R}) = \{M_g : g \in L^\infty(\mathbb{R})\}$$

as a von Neumann algebra, the closed linear span of

$$\{\widehat{T}_\beta \widehat{\psi} : \beta \in \mathbb{R}\}$$

is the set of square integrable functions on $\mathbb{R}$ with support $\subseteq$ supp$\widehat{\psi}$. Hence

$$\text{supp}((\widehat{T}_{-2^{-n}} - \widehat{T}^{-1}) \widehat{D}^n \widehat{T}^l \widehat{\psi})$$

must be disjoint (a.e.) from supp$(\widehat{\psi})$ for each n, l. We have $\widehat{T}_{-2^{-n}} = M_{e^{2^{-n} si}}$ and $\widehat{T}^{-1} = M_{e^{si}}$. For $n \neq 0$ the function $(e^{2^{-n} si} - e^{si})$ is nonzero (a.e.). So for $n \neq 0$

$$\text{supp}((\widehat{T}_{-2^{-n}} - \widehat{T}^{-1}) \widehat{D}^n \widehat{T}^l \widehat{\psi})$$

coincides with supp$(\widehat{D}^n \widehat{T}^l \widehat{\psi})$. Since $\widehat{D}^n \widehat{T}^l = \widehat{T}_{2^{-n}l} \widehat{D}^n$ and $\widehat{T}_{2^{-n}l} = M_{e^{-2^{-n} lsi}}$,

$$\text{supp}(\widehat{D}^n \widehat{T}^l \widehat{\psi}) = \text{supp}(\widehat{D}^n \widehat{\psi}).$$

Let $E = \text{supp}(\widehat{\psi})$. Then $\widehat{D}^n \widehat{\psi} = 2^{-\frac{n}{2}} \widehat{\psi}(2^{-n}(\cdot))$. So

$$\text{supp}(\widehat{D}^n \widehat{\psi}) = 2^n E.$$

So the above argument says that $2^n E \cap E = \emptyset$, $n \neq 0$. Thus $\{2^n E : n \in \mathbb{Z}\}$ is a disjoint family. Since $\{\widehat{D}^n \widehat{T}^l \widehat{\psi}\}$ is a basis for $L^2(\mathbb{R})$, the union of their supports must differ from $\mathbb{R}$ by at most a null set. So $\{2^n E : n \in \mathbb{Z}\}$ is a partition of $\mathbb{R}$.

Since $\{\widehat{D}^n \widehat{T}^l \widehat{\psi} : n, l \in \mathbb{Z}\}$ is an orthonormal basis for $L^2(\mathbb{R})$ and $\{2^n E : n \in \mathbb{Z}\}$ is a partition of $\mathbb{R}$, where $E = \text{supp}\widehat{\psi}$, it follows that $\{\widehat{T}^l \widehat{\psi} : l \in \mathbb{Z}\}$ is an o.n. basis for $L^2(E)$. Then Lemma 4.1 implies that

$$E \sim [0, 2\pi) \bmod 2\pi,$$

and

$$|\widehat{\psi}| = \frac{1}{\sqrt{2\pi}} \chi_E,$$

as required. $\square$

CHAPTER 5

Operator Interpolation of Wavelets

In this section we derive a method of *operator-theoretic interpolation* between certain special pairs of wavelets, and between single wavelets and special families of them. This generalizes Proposition 1.5 When applied to s-elementary wavelets, this yields a new construction of a class due to Meyer, and generalizes that class. While the s-elementary wavelets do not have "good" regularity properties because their Fourier transforms are discontinuous, they can be basic building blocks from which certain other wavelets with better regularity properties can be derived. It is hoped that further work here may yield synthesis of more general wavelets in terms of interpolation families of s-elementary wavelets.

If ψ, η are wavelets let $V := V_\psi^\eta$ be the (unique) unitary operator in $\mathcal{C}_\psi(D, T)$ with $V\psi = \eta$. Suppose that V *normalizes* $\{D, T\}'$ in the sense that

$$V^*(\{D, T\}')V = \{D, T\}'.$$

This will happen (5.3) if η and ψ are s-elementary. In this case the algebra, before closure, generated by $\{D, T\}'$ and V is the set of all finite sums (polynomials) of the form $\sum A_n V^n$, with coefficients $A_n \in \{D, T\}'$. The closure in the strong operator topology is a von Neumann algebra. If $V^n = I$ for some n this is $*$-isomorphic to the (finite-dimensional) *cross-product* ([**10**], [**14**]) of $\{D, T\}'$ under the automorphism group induced by V. This has a special matricial form. Now suppose further that *every power* of V is contained in $\mathcal{C}_\psi(D, T)$. This occurs only in special cases, yet it occurs frequently enough to yield some general methods. Then since $\mathcal{C}_\psi(D, T)$ is closed under left multiplication by $\{D, T\}'$ by Theorem 3.9(vi), this "cross-product" is contained in $\mathcal{C}_\psi(D, T)$, so its unitary group parameterizes a path-connected subset of $\mathcal{W}(D, T)$ that contains ψ and η via the correspondence $U \to U\psi$. We say that wavelets in this set are *interpolated* from $\{\psi, \eta\}$, and that $\{\psi, \eta\}$ *admits operator-interpolation*. More generally, if $\psi \in \mathcal{W}(D, T)$ is fixed and $\mathcal{F} \subseteq \mathcal{W}(D, T)$ is a family such that each $V_\psi^\eta, \eta \in \mathcal{F}$, normalizes $\{D, T\}'$ and

$$\text{Group}\{V_\psi^\eta : \eta \in \mathcal{F}\} \subset \mathcal{C}_\psi(D, T),$$

then if U is a unitary in the von Neumann algebra generated by $\{D, T\}'$ and $\{V_\psi^\eta : \eta \in \mathcal{F}\}$, we say that the wavelet $U\psi$ is interpolated from $\{\psi, \mathcal{F}\}$ and that $(\psi, \mathcal{F})$ admits operator-interpolation.

We follow with some basics of operator interpolation for s-elementary wavelets.

Let E and F be wavelet sets. Let $\sigma : E \to F$ be the 1-1, onto map implementing the 2π-translation congruence. Since E and F both generate partitions of $\mathbb{R} \setminus \{0\}$ under dilation by powers of 2, we may extend σ to a 1-1 map of $\mathbb{R}$ onto $\mathbb{R}$ by defining $\sigma(0) = 0$, and

$$\sigma(s) = 2^n \sigma(2^{-n}s) \quad \text{for} \quad s \in 2^n E, \ n \in \mathbb{Z}.$$

We adopt the notation σ_E^F for this, and call it the *interpolation map* for the ordered pair (E, F).

Lemma 5.1. *In the above notation, σ_E^F is a measure-preserving transformation from $\mathbb{R}$ onto $\mathbb{R}$.*

Proof. Let $\sigma := \sigma_E^F$. Let $\Omega \subseteq \mathbb{R}$ be a measurable set. Let $\Omega_n = \Omega \cap 2^n E$, $n \in \mathbb{Z}$, and let $E_n = 2^{-n}\Omega_n \subseteq E$. Then $\{\Omega_n\}$ is a partition of Ω, and we have $m(\sigma(E_n)) = m(E_n)$ because the restriction of σ to E is measure-preserving. So

$$
\begin{aligned}
m(\sigma(\Omega)) &= \sum_n m(\sigma(\Omega_n)) = \sum_n m(2^n \sigma(E_n)) \\
&= \sum_n 2^n m(\sigma(E_n)) = \sum_n 2^n m(E_n) \\
&= \sum_n m(2^n E_n) = \sum_n m(\Omega_n) = m(\Omega). \quad \square
\end{aligned}
$$

A function $f : \mathbb{R} \to \mathbb{R}$ is called 2-*homogeneous* if $f(2s) = 2f(s)$ for all $s \in \mathbb{R}$. (More generally a-homogeneous means $f(as) = af(s)$.) Equivalently, f is 2-homogeneous iff $f(2^n s) = 2^n f(s)$, $s \in \mathbb{R}, n \in \mathbb{Z}$. Such a function is completely determined by its values on any subset of $\mathbb{R}$ which generates a partition of $\mathbb{R} \setminus \{0\}$ by 2-dilation. So σ_E^F is the (unique) 2-homogeneous extension of the 2π-translation congruence $E \to F$. The set of all 2-homogeneous measure-preserving transformations of $\mathbb{R}$ clearly forms a group under composition. Also, the composition of a 2-dilation-periodic function with a 2-homogeneous function is clearly 2-dilation periodic. These facts will be useful.

If $\sigma : \mathbb{R} \to \mathbb{R}$ is a measure-preserving transformation, let U_σ denote the unitary composition operator defined by

$$
U_\sigma f = f \circ \sigma^{-1}, f \in L^2(\mathbb{R}).
$$

We write $U_E^F := U_{\sigma_E^F}$. Clearly $(\sigma_E^F)^{-1} = \sigma_F^E$ and $(U_E^F)^* = U_F^E$. We have $U_E^F \widehat{\psi}_E = \widehat{\psi}_F$ since $\sigma_E^F(E) = F$.

Theorem 5.2. *Let E and F be wavelet sets. Then:*
1. $U_E^F \in \mathcal{C}_{\widehat{\psi}_E}(\widehat{D}, \widehat{T})$;
2. U_E^F *normalizes* $\{\widehat{D}, \widehat{T}\}'$;
3. *If* $E \neq F$ *then* $U_E^F \notin \{\widehat{D}, \widehat{T}\}''$.

Proof. (i) Write $\sigma = \sigma_E^F$ and $U_\sigma = U_E^F$. We have $U_\sigma \widehat{\psi}_E = \widehat{\psi}_F$ since $\sigma(E) = F$. We must show

$$
U_\sigma \widehat{D}^n \widehat{T}^l \widehat{\psi}_E = \widehat{D}^n \widehat{T}^l U_\sigma \widehat{\psi}_E, \; n, l \in \mathbb{Z}.
$$

We have

$$
\begin{aligned}
(U_\sigma \widehat{D}^n \widehat{T}^l \widehat{\psi}_E)(s) &= (U_\sigma \widehat{D}^n e^{-ils} \widehat{\psi}_E)(s) \\
&= U_\sigma 2^{-\frac{n}{2}} e^{-il2^{-n}s} \widehat{\psi}_E(2^{-n}s) \\
&= 2^{-\frac{n}{2}} e^{-il2^{-n}\sigma^{-1}(s)} \widehat{\psi}_E(2^{-n}\sigma^{-1}(s)).
\end{aligned}
$$

Note that $\chi_E(2^{-n}\sigma^{-1}(s)) = 1$ iff $2^{-n}\sigma^{-1}(s) \in E$ iff $s \in \sigma(2^n E) = 2^n\sigma(E) = 2^n F$ iff $2^{-n}s \in F$. Also, if $2^{-n}s \in F$, then $2^{-n}s - \sigma^{-1}(2^{-n}s) = 2^{-n}s - 2^{-n}\sigma^{-1}(s) \in 2\mathbb{Z}\pi$. Thus $e^{-il2^{-n}\sigma^{-1}(s)}\widehat{\psi}_E(2^{-n}\sigma^{-1}(s)) = e^{-il2^{-n}s}\widehat{\psi}_F(2^{-n}s)$. Also, we have

$$(\widehat{D}^n\widehat{T}^l U_\sigma \widehat{\psi}_E)(s) = (\widehat{D}^n\widehat{T}^l\widehat{\psi}_F)(s) = 2^{-\frac{n}{2}}e^{-il2^{-n}s}\widehat{\psi}_F(2^{-n}s).$$

(ii) By Theorem 3.5 the generic element of $\{\widehat{D}, \widehat{T}\}'$ has the form M_h for some 2-dilation periodic function $h \in L^\infty(\mathbb{R})$. Since $U_E^F = U_\sigma$ for $\sigma = \sigma_E^F$, we have $U_E^F M_h U_F^E = M_{h\circ\sigma^{-1}}$. We have

$$(h \circ \sigma^{-1})(2s) = h(\sigma^{-1}(2s)) = h(2\sigma^{-1}(s)) = h(\sigma^{-1}(s)), \quad s \in \mathbb{R},$$

so $h \circ \sigma^{-1}$ is 2-dilation periodic, completing the proof.

(iii) Since $E \neq F$, σ_E^F is not (a.e.) the identity map. So for some set $E_0 \subset E$ of positive measure and some nonzero integer n we have

$$\sigma_E^F(s) = s + 2\pi n, \quad s \in E_0.$$

Suppose by way of contradiction that $U_E^F \in \{\widehat{D}, \widehat{T}\}''$. Then for *each* 2-dilation periodic function $h \in L^\infty(\mathbb{R})$ we have

$$U_E^F M_h = M_h U_E^F.$$

So $M_h = U_E^F M_h U_F^E = M_{h\circ\sigma^{-1}}$, where $\sigma = \sigma_E^F$. Since $h = h \circ \sigma^{-1}$ for each bounded 2-dilation periodic function h, it follows that $\frac{\sigma^{-1}(s)}{s}$ is an integral multiple of 2 for each $s \neq 0$ (a.e.). Hence $\frac{\sigma(s)}{s}$ is an integral multiple of 2 (a.e.). It follows further, then, that there is a measurable subset $E_1 \subset E_0$ of positive measure, and an integer k, such that $\sigma(s) = 2^k s$ for all $s \in E_1$. Then $2^k s = s + 2\pi n$ for all $s \in E_1$. Since $n \neq 0$, we must have $k \neq 0$. But this forces E_1 to be a singleton (hence null) set, a contradiction. $\square$

Corollary 5.3. $\mathcal{C}_\psi(D, T)$ is nonabelian for every $\psi \in \mathcal{W}(D, T)$.

Proof. Part (iii) of Theorem 5.2 shows that $\mathcal{C}_{\widehat{\psi}_E}(\widehat{D}, \widehat{T})$ is nonabelian, for instance, for $E = [-2\pi, -\pi) \cup [\pi, 2\pi)$. The result then follows from Proposition 1.8. $\square$

The above theorem shows that a pair (E, F) of wavelet sets (or, rather, their corresponding s-elementary wavelets) admits operator-interpolation if and only if Group$\{U_E^F\}$ is contained in the local commutant $\mathcal{C}_{\widehat{\psi}_E}(\widehat{D}, \widehat{T})$, since the requirement that U_E^F normalizes $\{\widehat{D}, \widehat{T}\}'$ is automatically satisfied. It is easy to see that this is equivalent to the condition that for each $n \in \mathbb{Z}$, σ^n is a 2π-congruence of E in the sense that $(\sigma^n(s) - s)/2\pi \in \mathbb{Z}$ for all $s \in E$, which in turn implies that $\sigma^n(E)$ is a wavelet set for all n. Here $\sigma = \sigma_E^F$. This property holds trivially if σ is *involutive* (i.e. $\sigma^2 =$ identity).

In cases where "torsion" is present, so $(\sigma_E^F)^k$ is the identity map for some finite integer k, the von Neumann algebra generated by $\{\widehat{D}, \widehat{T}\}'$ and $U := U_E^F$ has the simple form

$$\{\sum_{n=0}^{k} M_{h_n} U^n : h_n \in L^\infty(\mathbb{R}) \text{ with } h_n(2s) = h_n(s), \ s \in \mathbb{R}\},$$

and so each member of this "interpolated" family of wavelets has the form

$$\frac{1}{\sqrt{2\pi}} \sum_0^k h_n(s) \chi_{\sigma^n(E)}$$

for 2-dilation periodic "coefficient" functions $\{h_n(s)\}$ which satisfy the necessary and sufficient condition that the operator

$$\sum_{n=0}^k M_{h_n} U^n$$

is unitary.

A standard computation shows that the map θ sending $\sum_0^k M_{h_n} U^n$ to the $k \times k$ function matrix (h_{ij}) given by

$$h_{ij} = h_{\alpha(i,j)} \circ \sigma^{-(i-1)}$$

where $\alpha(i,j) = (i+1)$ modulo k, is a $*$-isomorphism. (This matricial algebra is the cross-product of $\{\widehat{D}, \widehat{T}\}'$ by $\text{ad}(U_E^F)$.) So, for instance, if $k = 3$ then θ maps

$$M_{h_1} + M_{h_2} U_E^F + M_{h_3}(U_E^F)^2$$

to

$$\begin{pmatrix} h_1 & h_2 & h_3 \\ h_3 \circ \sigma^{-1} & h_1 \circ \sigma^{-1} & h_2 \circ \sigma^{-1} \\ h_2 \circ \sigma^{-2} & h_3 \circ \sigma^{-2} & h_1 \circ \sigma^{-2} \end{pmatrix}.$$

This shows that $\sum_0^k M_{h_n} U^n$ is a unitary operator iff the scalar matrix $(h_{ij})(s)$ is unitary for almost all $s \in \mathbb{R}$. This *Coefficient Criterion* yields interpolation formulas.

The involutive case seems to be common. See examples A.1, A.4 and A.8 in the Appendix. Example A.4 shows that *uncountable commutative groups* of involutive interpolation maps exist. Example A.9 shows that the case $\sigma^3 = $ identity is possible. *Question:* Is the case $\sigma^n = $ identity attainable for each positive integer n? Let us say that a pair of wavelet sets (E, F) is an *interpolation pair* if $(\sigma_E^F)^2 = $ identity. In this case $\sigma_E^F = \sigma_F^E$. More generally, it is natural to define an *interpolation family* of wavelet sets (based at E) to be a family $\mathcal{E}$ of wavelet sets, "based" at a single special fixed wavlet set $E \in \mathcal{E}$, with the property that

$$\{\sigma_E^F : F \in \mathcal{E}\}$$

is a group. For a *finite* interpolation family the matricial cross-product form for the von Neumann algebra generated by $\{D, T\}'$ together with the corresponding group of interpolation unitaries will yield interpolation formulas. (For an infinite family the von Neumann algebra may not be $*$-isomorphic to the cross-product.)

Problem E. *Characterize those groups which are isomorphic to groups of interpolation maps for interpolation families of wavelet sets.*

The Coefficient Criterion for the case $k = 2$ yields:

Proposition 5.4. *If (E, F) is an interpolation pair then*

$$\widehat{\psi}(s) = h_1(s)\widehat{\psi}_E(s) + h_2(s)\widehat{\psi}_F(s) \tag{$*$}$$

is the Fourier Transform of an orthogonal wavelet whenever h_1 *and* h_2 *are 2-dilation-periodic functions on* $\mathbb{R}$ *with the property that the matrix*

$$\begin{pmatrix} h_1 & h_2 \\ h_2 \circ \sigma_E^F & h_1 \circ \sigma_E^F \end{pmatrix} \qquad (**)$$

is unitary (a.e.).

We shall show that Meyer's (family of) wavelets have the above form. Meyer's class is (c.f. [**8**],p.117):

$$\widehat{\psi}_{Me}(s) = \begin{cases} \frac{1}{\sqrt{2\pi}} e^{\frac{is}{2}} \cos[\frac{\pi}{2}\nu(-\frac{3s}{4\pi} - 1)], & s \in [-\frac{8\pi}{3}, -\frac{4\pi}{3}), \\ \frac{1}{\sqrt{2\pi}} e^{\frac{is}{2}} \sin[\frac{\pi}{2}\nu(-\frac{3s}{2\pi} - 1)], & s \in [-\frac{4\pi}{3}, -\frac{2\pi}{3}) \\ \frac{1}{\sqrt{2\pi}} e^{\frac{is}{2}} \sin[\frac{\pi}{2}\nu(\frac{3s}{2\pi} - 1)], & s \in [\frac{2\pi}{3}, \frac{4\pi}{3}) \\ \frac{1}{\sqrt{2\pi}} e^{\frac{is}{2}} \cos[\frac{\pi}{2}\nu(\frac{3s}{4\pi} - 1)], & s \in [\frac{4\pi}{3}, \frac{8\pi}{3}) \\ 0 & \text{otherwise} \end{cases}$$

for $s \in \mathbb{R}$, where ν is a real-valued function which satisfies the relation

$$\nu(s) + \nu(1 - s) = 1, \ s \in \mathbb{R}.$$

Normally, one chooses ν so that $\widehat{\psi}_{Me}$ has desired regularity properties. If ν is taken with $\nu(s) = 0$ for $s \leq 0$ and $\nu(s) = 1$ for $s \geq 1$, then if ν is continuous, or in class $\mathcal{C}^k$, or $\mathcal{C}^\infty$, then the function $\widehat{\psi}_{Me}$ is in the same class. Any choice of a measurable real valued function ν satisfying $\nu(s) + \nu(1 - s) = 1$ yields a (perhaps "badly behaved") wavelet, however.

Proposition 5.5. *The wavelets* $\widehat{\psi}_{Me}$ *have the interpolation form* (*).

Proof. Decompose the support of $\widehat{\psi}_{Me}$ as $E \cup F$, where $E = [-\frac{8\pi}{3}, -\frac{4\pi}{3}) \cup [\frac{2\pi}{3}, \frac{4\pi}{3})$ and $F = [-\frac{4\pi}{3}, -\frac{2\pi}{3}) \cup [\frac{4\pi}{3}, \frac{8\pi}{3})$. These are the wavelet sets $E_{-\frac{\pi}{3}}$ and $E_{\frac{\pi}{3}}$, respectively, from 4.5(ii). Then $(\sigma_E^F)^2 = $ identity, as shown in A.1. For this degenerate case we have

$$\sigma_E^F(s) = \begin{cases} s + 4\pi, & s \in [-\frac{8\pi}{3}, -\frac{4\pi}{3}) \\ s - 2\pi, & s \in [\frac{2\pi}{3}, \frac{4\pi}{3}) \end{cases}$$

Let

$$E_- = [-\frac{8\pi}{3}, -\frac{4\pi}{3}), E_+ = [\frac{2\pi}{3}, \frac{4\pi}{3}), F_- = [-\frac{4\pi}{3}, -\frac{2\pi}{3}), F_+ = [\frac{4\pi}{3}, \frac{8\pi}{3}).$$

Then

$$\widehat{\psi}_{Me} = f\widehat{\psi}_E + g\widehat{\psi}_F,$$

with f defined on E by

$$f(s) = \begin{cases} e^{\frac{is}{2}} \cos[\frac{\pi}{2}\nu(-\frac{3s}{4\pi} - 1)], & s \in E_-, \\ e^{\frac{is}{2}} \sin[\frac{\pi}{2}\nu(\frac{3s}{2\pi} - 1)], & s \in E_+ \end{cases}$$

and extended to $\mathbb{R}$ 2-dilation periodically by setting $f(0) = 0$ and $f(s) = f(2^{-n}s)$ for $s \in 2^n E$, $n \in \mathbb{Z}$, and similarly with g defined on $\mathbb{R}$ by

$$g(s) = \begin{cases} e^{\frac{is}{2}} \sin[\frac{\pi}{2}\nu(-\frac{3s}{2\pi} - 1)], & s \in F_- \\ e^{\frac{is}{2}} \cos[\frac{\pi}{2}\nu(\frac{3s}{4\pi} - 1)], & s \in F_+ \end{cases}$$

and $g(0) = 1$, $g(s) = g(2^{-n}s)$ for $s \in 2^n F$, $n \in \mathbb{Z}$. With this extension

$$f(s) = \begin{cases} e^{is} \cos[\frac{\pi}{2}\nu(-\frac{3s}{2\pi} - 1)], & s \in F_-, \\ e^{\frac{is}{4}} \sin[\frac{\pi}{2}\nu(\frac{3s}{4\pi} - 1)], & s \in F_+, \end{cases}$$

and

$$g(s) = \begin{cases} e^{\frac{is}{4}} \sin[\frac{\pi}{2}\nu(-\frac{3s}{4\pi} - 1)], & s \in E_-, \\ e^{is} \cos[\frac{\pi}{2}\nu(\frac{3s}{2\pi} - 1)], & s \in E_+. \end{cases}$$

We must show that (a.e.) the matrix

$$\begin{pmatrix} f(s) & g(s) \\ g(\sigma(s)) & f(\sigma(s)) \end{pmatrix}$$

is unitary, $s \in \mathbb{R}$, where $\sigma = \sigma_E^F$. It will suffice to verify this for $s \in E$. We have $\sigma(E_-) = F_+$ and $\sigma(E_+) = F_-$. So on E_-,

$$\begin{aligned}
(f \circ \sigma)(s) = f(s + 4\pi) &= e^{\frac{i(s+4\pi)}{4}} \sin[\frac{\pi}{2}\nu(\frac{3(s+4\pi)}{4\pi} - 1)] \\
&= -e^{\frac{is}{4}} \sin[\frac{\pi}{2}\nu(\frac{3s}{4\pi} + 2)] \\
&= -e^{\frac{is}{4}} \sin[\frac{\pi}{2}(1 - \nu(-\frac{3s}{4\pi} - 1))] \\
&= -e^{\frac{is}{4}} \cos[\frac{\pi}{2}\nu(-\frac{3s}{4\pi} - 1)],
\end{aligned}$$

where we use the fact that $\nu(x) + \nu(1 - x) = 1$, $x \in \mathbb{R}$.
And on E_+,

$$\begin{aligned}
(f \circ \sigma)(s) = f(s - 2\pi) &= e^{i(s-2\pi)} \cos[\frac{\pi}{2}\nu(-\frac{3(s - 2\pi)}{2\pi} - 1)] \\
&= e^{is} \cos[\frac{\pi}{2}\nu(-\frac{3s}{2\pi} + 2)] \\
&= e^{is} \sin[\frac{\pi}{2}\nu(\frac{3s}{2\pi} - 1)].
\end{aligned}$$

Similarly,

$$(g \circ \sigma)(s) = \begin{cases} e^{\frac{is}{2}} \sin[\frac{\pi}{2}\nu(-\frac{3s}{4\pi} - 1)], & s \in E_- \\ -e^{\frac{is}{2}} \cos[\frac{\pi}{2}\nu(\frac{3s}{2\pi} - 1)], & s \in E_+. \end{cases}$$

For convenience, let $\gamma_1 = \frac{\pi}{2}\nu(-\frac{3s}{4\pi} - 1)$ and $\gamma_2 = \frac{\pi}{2}\nu(\frac{3s}{2\pi} - 1)$. Then for $s \in E_-$ we have

$$\begin{aligned}
\begin{pmatrix} f(s) & g(s) \\ g(\sigma(s)) & f(\sigma(s)) \end{pmatrix} &= \begin{pmatrix} e^{\frac{is}{2}} \cos\gamma_1 & e^{\frac{is}{4}} \sin\gamma_1 \\ e^{\frac{is}{2}} \sin\gamma_1 & -e^{\frac{is}{4}} \cos\gamma_1 \end{pmatrix} \\
&= \begin{pmatrix} \cos\gamma_1 & \sin\gamma_1 \\ \sin\gamma_1 & -\cos\gamma_1 \end{pmatrix} \cdot \begin{pmatrix} e^{\frac{is}{2}} & 0 \\ 0 & e^{\frac{is}{4}} \end{pmatrix}
\end{aligned}$$

a product of two unitaries, hence unitary. Similarly, for $s \in E_+$ we have

$$\begin{aligned}
\begin{pmatrix} f(s) & g(s) \\ g(\sigma(s)) & f(\sigma(s)) \end{pmatrix} &= \begin{pmatrix} e^{\frac{is}{2}} \sin\gamma_2 & e^{is} \cos\gamma_2 \\ -e^{\frac{is}{2}} \cos\gamma_2 & e^{is} \sin\gamma_2 \end{pmatrix} \\
&= \begin{pmatrix} \sin\gamma_2 & \cos\gamma_2 \\ -\cos\gamma_2 & \sin\gamma_2 \end{pmatrix} \cdot \begin{pmatrix} e^{\frac{is}{2}} & 0 \\ 0 & e^{is} \end{pmatrix}.
\end{aligned}$$

Again each factor is unitary. The proof is complete. $\square$

We note that we have proven that ψ_{Me} satisfies the definition of an orthogonal wavelet without carrying out sensitive integral identities. No regularity hypothesis on ν are needed for this.

Proposition 5.5 shows that Meyer's wavelet can be obtained by interpolation between a pair of wavelet sets by showing that it satisfies our "coefficient criterion." But it does not give algorithms, or give insight into the special form of $\widehat{\psi}_{Me}$. We now address these matters. (See especially Example 5.13.)

Let (E, F) be an interpolation pair. The reason that the matrix criterion $(**)$ does not automatically yield algorithms is that the condition involves composition with σ_E^F. We have

$$\text{w}^*(\{\widehat{D}, \widehat{T}\}', U_E^F) \subseteq \mathcal{C}_{\widehat{\psi}_E}(\widehat{D}, \widehat{T}),$$

so the unitary group of this *interpolation von Neumann algebra* (denote it by $\mathcal{B}(E, F)$) parameterizes the interpolated wavelets. We consider a special abelian subgroup of this group which is particularly easy to parameterize. Let

$$\mathcal{B}_0(E, F) = \mathcal{B}(E, F) \cap (U_E^F)'.$$

Then

$$\mathcal{B}_0(E, F) = \text{w}^*(\{\widehat{D}, \widehat{T}\}' \cap \{U_E^F\}', U_E^F).$$

Theorem 5.2 (iii) shows that $\mathcal{B}(E, F)$ is always nonabelian if $E \neq F$, so the inclusion $\mathcal{B}_0(E, F) \subseteq \mathcal{B}(E, F)$ is always proper.

Lemma 5.6. *If $h \in L^\infty$ and if σ is a measure-preserving transformation of $\mathbb{R}$ onto $\mathbb{R}$, then $U_\sigma M_h = M_h U_\sigma$ if and only if $h = h \circ \sigma$.*

Proof. If $f \in L^2(\mathbb{R})$, then $(U_\sigma M_h f)(s) = h(\sigma^{-1}(s))f(\sigma^{-1}(s))$ and $(M_h U_\sigma f)(s) = h(s)f(\sigma^{-1}(s))$. These are equal for all f iff $h = h \circ \sigma^{-1}$ iff $h = h \circ \sigma$. $\square$

Since a matrix

$$\begin{pmatrix} a & b \\ b & a \end{pmatrix}, \ a, b \in \mathbb{C}$$

is unitary iff both

$$|a + b| = 1 \text{ and } |a - b| = 1,$$

for the abelian case condition $(**)$ reduces to the condition that both

$$(h_1 + h_2) \text{ and } (h_1 - h_2)$$

are unimodular. So we have an algorithm:

Proposition 5.7. *Let (E, F) be an interpolation pair of wavelet sets, and let f and g be arbitrary 2-dilation- periodic unimodular functions on $\mathbb{R}$ with $f \circ \sigma_E^F = f$ and $g \circ \sigma_E^F = g$. Then*

$$\widehat{\psi} = (\frac{f + g}{2})\widehat{\psi}_E + (\frac{f - g}{2})\widehat{\psi}_F \qquad (* * *)$$

is the Fourier transform of a wavelet.

Proof. Apply the above discussion with $h_1 = (f + g)/2$ and $h_2 = (f - g)/2$. $\square$

Remarks 5.8. (i) Note that Proposition 5.7 generalizes Proposition 1.5 in the special case of wavelets. The form $(* * *)$ is clearly equivalent to the form

$$\widehat{\psi}(s) = e^{i\alpha(s)}[\cos \beta(s)\widehat{\psi}_E + i \sin \beta(s)\widehat{\psi}_F]$$

for α, β arbitrary real-valued measurable 2-dilation-periodic functions with

$$\alpha \circ \sigma_E^F = \alpha \quad \text{and} \quad \beta \circ \sigma_E^F = \beta.$$

(ii) If we drop the requirement in part (i) that $\alpha \circ \sigma_E^F = \alpha$, and only require that α be 2-dilation-periodic, then $\widehat{\psi}$ is still the Fourier transform of a wavelet. This is because *arbitrary* 2-dilation-periodic unimodular functions are wavelet multipliers in the sense of Remark 3.7. The requirement on β cannot be dropped, however.

By construction, the coefficient criterion $(\ast\ast)$ is invariant under 2-dilation-periodic wavelet multipliers. We next show it is also invariant (in the only appropriate sense possible) under 2π-translation-periodic wavelet multipliers.

Proposition 5.9. *Let (E_1, E_2) be an interpolation pair of wavelet sets, and let h_1, h_2 be 2-dilation-periodic functions on $\mathbb{R}$ such that*

$$\begin{pmatrix} h_1(s) & h_2(s) \\ h_2(\sigma(s)) & h_1(\sigma(s)) \end{pmatrix}$$

is unitary (a.e.) on $\mathbb{R}$, where $\sigma := \sigma_E^F$. Let f be an arbitrary 2π-translation-periodic unimodular function.

Define 2-dilation-periodic functions f_1, f_2 on $\mathbb{R}$ by setting $f_i = f|_{E_i}$ and extending 2-dilation-periodically to $\mathbb{R} \setminus \{0\}$, and setting $f_i(0) = 1$. Let $\widetilde{h}_i(s) = h_i(s)f_i(s)$, $s \in \mathbb{R}$, $i = 1, 2$. Then

$$\begin{pmatrix} \widetilde{h}_1(s) & \widetilde{h}_2(s) \\ \widetilde{h}_2(\sigma(s)) & \widetilde{h}_1(\sigma(s)) \end{pmatrix}$$

is unitary (a.e.) on $\mathbb{R}$.

Proof. Since $\widetilde{h}_1, \widetilde{h}_2$ are 2-dilation periodic it is only necessary to verify that the last matrix is unitary (a.e.) on E_1. For $s \in E_1$, let $k(s)$ be the unique integer such that $2^{k(s)}s \in E_2$. Then $2^{k(s)}\sigma(s) = \sigma(2^{k(s)}s) \in E_1$. We have $f_1(s) = f(s)$, and $f_2(s) = f_2(2^{k(s)}s) = f(2^{k(s)}s)$. Also,

$$f_1(\sigma(s)) = f_1(2^{k(s)}\sigma(s)) = f_1(\sigma(2^{k(s)}s)) = f(\sigma(2^{k(s)}s)) = f(2^{k(s)}s),$$

where we used dilation-periodicity of f_1, translation-periodicity of f, 2-homogeneity of σ, and the fact that $f_1|_{E_1} = f|_{E_1}$. In addition,

$$f_2(\sigma(s)) = f(\sigma(s)) = f(s),$$

since $f_2|_{E_2} = f|_{E_2}$ and σ acts as a 2π-congruence on E_2. Hence for (a.e.) $s \in E_1$,

$$\begin{pmatrix} \widetilde{h}_1(s) & \widetilde{h}_2(s) \\ \widetilde{h}_2(\sigma(s)) & \widetilde{h}_1(\sigma(s)) \end{pmatrix} = \begin{pmatrix} h_1(s)f(s) & h_2(s)f(2^{k(s)}s) \\ h_2(s)f(s) & h_1(s)f(2^{k(s)}s) \end{pmatrix}$$

$$= \begin{pmatrix} h_1(s) & h_2(s) \\ h_2(\sigma(s)) & h_1(\sigma(s)) \end{pmatrix} \cdot \begin{pmatrix} f(s) & 0 \\ 0 & f(2^{k(s)}s) \end{pmatrix},$$

a product of two unitaries, hence unitary. $\square$

Sometimes parts of two wavelet sets can be combined to form a third wavelet set.

Proposition 5.10. *Let (E, F) be an interpolation pair, and let $G \subseteq \mathbb{R}$ be a measurable set which is 2-dilation stable in the sense that $2G = G$ and which is invariant under σ_E^F. Then*

$$(G \cap E) \cup (G^c \cap F)$$

is also a wavelet set.

Proof. Let

$$h_1 = \chi_G \quad \text{and} \quad h_2 = \chi_{G^c} = 1 - \chi_G.$$

The hypotheses imply that h_i is 2-dilation periodic and $h_i \circ \sigma_E^F = h_i$, $i = 1, 2$. Since $(h_1 + h_2)$ and $(h_1 - h_2)$ are unimodular, an application of $(**)$ or Proposition 5.7 implies that

$$\widehat{\psi} := \frac{1}{\sqrt{2\pi}} h_1 \chi_E + \frac{1}{\sqrt{2\pi}} h_2 \chi_F$$

is the Fourier transform of a wavelet. Since $\widehat{\psi} = \frac{1}{\sqrt{2\pi}} \chi_K$, where $K = (G \cap E) \cup (G^c \cap F)$, K must be a wavelet set. $\square$

We next show that there are always nontrivial f and g satisfying the hypothesis of Proposition 5.7. The proof is in fact a construction.

Theorem 5.11. *Let (E, F) be an interpolation pair. Let*

$$\Lambda(E, F) = \{s \in E : \sigma_E^F(s) \in 2^n E \text{ for some nonnegative integer } n.\}.$$

Then $\Lambda(E, F)$ has positive measure. If $h(s)$ is an arbitrary bounded measurable function on $\Lambda(E, F)$, then h extends uniquely to a 2-dilation periodic function (which we still denote by h) on $\mathbb{R}$ which satisfies the condition that $h \circ \sigma_E^F = h$.

Proof. For each $n \in \mathbb{Z}$ let

$$E_n = \{s \in E : \sigma_E^F(s) \in 2^n E\}.$$

Then $\{E_n : n \in \mathbb{Z}\}$ is a partition of E. We have

$$\Lambda(E, F) = \cup\{E_n : n \geq 0\}.$$

If $s \in E_n$, let $\widetilde{s} = 2^{-n} \sigma_E^F(s) \in E$. Then

$$\sigma_E^F(\widetilde{s}) = \sigma_E^F(2^{-n} \sigma_E^F(s)) = 2^{-n} s,$$

so $\widetilde{s} \in E_{-n}$. This argument is reversible. So

$$2^{-n} \sigma_E^F(E_n) = E_{-n}, \ n \in \mathbb{Z}.$$

Suppose $\Lambda(E, F)$ is null. Then E_n is a null set for $n \geq 0$. So $\sigma_E^F(E_n)$ is null since σ_E^F is measure-preserving. Thus

$$E_{-n} = 2^{-n} \sigma_E^F(E_n)$$

is null. But $\cup_{n \in \mathbb{Z}} E_n = E$ and $m(E) = 2\pi$, a contradiction. Hence $\Lambda(E, F)$ has positive measure.

Now let $h(s)$ be an arbitrary bounded measurable function on $\Lambda(E, F)$. First extend to E by setting, for $s \in E_{-n}$ with $n > 0$,

$$h(s) = h(2^n \sigma_E^F(s)).$$

Then extend to $\mathbb{R}$ 2-dilation periodically by setting $h(s) = h(2^{-l}s)$ for $s \in 2^l E$. We claim $h = h \circ \sigma_E^F$. If $s \in E_n$ with $n > 0$ then $2^{-n}\sigma_E^F(s) \in E_{-n}$. So by 2-dilation periodicity of h and the definition of the extension to E_{-n},

$$h(\sigma_E^F(s)) = h(2^{-n}\sigma_E^F(s)) = h(2^n\sigma_E^F(2^{-n}\sigma_E^F(s))) = h(s),$$

where we use

$$\sigma_E^F(2^{-n}\sigma_E^F(s)) = 2^{-n}\sigma_E^F(\sigma_E^F(s)) = 2^{-n}s.$$

For $s \in E_{-n}$ with $n > 0$, we have

$$2^n\sigma_E^F(s) \in E_n.$$

So

$$\begin{aligned} h(\sigma_E^F(s)) &= h(2^n\sigma_E^F(s)) = h(\sigma_E^F(2^n\sigma_E^F(s))) \\ &= h(2^n\sigma_E^F(\sigma_E^F(s))) = h(2^n s) = h(s). \end{aligned}$$

Note that $E_0 = E \cap F$. So for $s \in E_0$, $\sigma_E^F(s) = s$. Thus the result for this case is trivial. $\square$

Corollary 5.12. *Let (E, F) be an interpolation pair of wavelet sets, and let ψ be a wavelet obtained by the method of operator-interpolation (∗) between ψ_E and ψ_F. If $\mathrm{supp}(\widehat{\psi})$ is not a wavelet set, then $\{ D, T, P_\psi \}'$ lies strictly between $\mathbb{C}I$ and $\{D, T\}'$.*

Proof. We shall use the characterization of $\{ D, T, P_\psi \}'$ given by Corollary 3.15. We have $\mathrm{supp}(\widehat{\psi}) \subseteq E \cup F$. Let $\Lambda(E, F)$ be as in Theorem 5.11, let $h \in L^\infty(\Lambda(E, F))$ be arbitrary, with the 2-dilation extension given by the proposition satisfying $h \circ \sigma_E^F = h$. We claim that h coincides on $E \cup F$ with a 2π-periodic function. It will be sufficient to show that if $s \in E \cup F$, and if $s + 2\pi n \in E \cup F$ for some $n \in \mathbb{Z}$, then $h(s + 2\pi n) = h(s)$. If $s \in E$, then $s + 2\pi n \in E \cup F$ implies $s + 2\pi n \in F$ and $\sigma_E^F = s + 2\pi n$. Hence $h(s + 2\pi n) = h(\sigma_E^F(s)) = h(s)$. The argument for $s \in F$ is similar. Since $\Lambda(E, F)$ has positive measure, this shows that $\{D, T, P_\psi\}'$ is nontrivial. If $\{D, T, P_\psi\}' = \{D, T\}'$, then by Corollary 3.15, every 2-dilation-periodic function coincides on $\mathrm{supp}(\widehat{\psi})$ with a 2π-translation-periodic function. This implies $\mathrm{supp}(\widehat{\psi})$ is contained in a wavelet set, so since ψ is a wavelet, it must fill out the wavelet set. Thus if $\mathrm{supp}(\widehat{\psi})$ is not a wavelet set, then $\{D, T, P_\psi\}' \neq \{D, T\}'$. $\square$

Example 5.13. We give an example which demonstrates how wavelets with special properties, such as $\widehat{\psi}_{Me}$, can be constructed *algebraically* from basic elements of the interpolation theory without resorting to integral identities. Let

$$E = [-\frac{8\pi}{3}, -\frac{4\pi}{3}) \cup [\frac{2\pi}{3}, \frac{4\pi}{3}) = E_- \cup E_+$$

and

$$F = [-\frac{4\pi}{3}, -\frac{2\pi}{3}) \cup [\frac{4\pi}{3}, \frac{8\pi}{3}) = F_- \cup F_+$$

as in the proof of Proposition 5.5. Then (E, F) is an interpolation pair, and

$$\sigma_E^F(E_-) = E_- + 4\pi = F_+, \sigma_E^F(E_+) = E_+ - 2\pi = F_-.$$

We have $\Lambda(E, F) = [-\frac{8\pi}{3}, -\frac{4\pi}{3})$. Then Theorem 5.11 and Proposition 5.7 imply that if f and g are *arbitrary unimodular functions* defined on $[-\frac{8\pi}{3}, -\frac{4\pi}{3})$, then $\widehat{\eta}_{f,g}$ is (the Fourier transform of) a wavelet, where

$$\widehat{\eta}_{f,g}(s) = \begin{cases} \frac{1}{\sqrt{2\pi}}\left(\frac{f(s)+g(s)}{2}\right), & s \in [-\frac{8\pi}{3}, -\frac{4\pi}{3}) \\ \frac{1}{\sqrt{2\pi}}\left(\frac{f(2s)-g(2s)}{2}\right), & s \in [-\frac{4\pi}{3}, -\frac{2\pi}{3}) \\ \frac{1}{\sqrt{2\pi}}\left(\frac{f(2s-4\pi)+g(2s-4\pi)}{2}\right), & s \in [\frac{2\pi}{3}, \frac{4\pi}{3}) \\ \frac{1}{\sqrt{2\pi}}\left(\frac{f(s-4\pi)-g(s-4\pi)}{2}\right), & s \in [\frac{4\pi}{3}, \frac{8\pi}{3}) \\ 0, & \text{otherwise} \end{cases}$$

So suppose $\gamma(s)$ is an *arbitrary* measurable real-valued function on $[-\frac{8\pi}{3}, -\frac{4\pi}{3})$. Let $f = e^{i\gamma}, g = e^{-i\gamma}$. Write $\widehat{\eta}_{\gamma} := \widehat{\eta}_{f,g}$. Then

$$\widehat{\eta}_{\gamma}(s) = \begin{cases} \frac{1}{\sqrt{2\pi}}\cos\gamma(s), & s \in [-\frac{8\pi}{3}, -\frac{4\pi}{3}) \\ \frac{i}{\sqrt{2\pi}}\sin\gamma(2s), & s \in [-\frac{4\pi}{3}, -\frac{2\pi}{3}) \\ \frac{1}{\sqrt{2\pi}}\cos\gamma(2s-4\pi), & s \in [\frac{2\pi}{3}, \frac{4\pi}{3}) \\ \frac{i}{\sqrt{2\pi}}\sin\gamma(s-4\pi), & s \in [\frac{4\pi}{3}, \frac{8\pi}{3}) \\ 0, & \text{otherwise.} \end{cases}$$

Note that $\widehat{\eta}_{\gamma}$ cannot be continuous, for continuity would imply

$$\lim_{s\to(-\frac{4\pi}{3})^-} \cos\gamma(-\frac{4\pi}{3}) = i\sin\gamma(-\frac{8\pi}{3}).$$

Since $\gamma(s)$ is real this implies $\sin\gamma(-\frac{8\pi}{3}) = 0$; but also $\widehat{\eta}_{\gamma}(-\frac{8\pi}{3}) = 0$ implies $\cos\gamma(-\frac{8\pi}{3}) = 0$. Hence $\cos\gamma(-\frac{8\pi}{3}) = 0 = \sin\gamma(-\frac{8\pi}{3})$, an impossibility.

Now define $k(s) := i$ for $s \in E_-$, and $k(s) := 1$ for $s \in E_+$, and extend 2π-periodically to $\mathbb{R}$. Define $h(s) := -i$ for $s < 0$ and $h(s) := -1$ for $s \geq 0$, a 2-dilation periodic unimodular function. Since h, k are wavelet multipliers, $hk\widehat{\eta}_{\gamma}$ is the Fourier transform of a wavelet. On $E \cup F$,

$$k(s)h(s) = \begin{cases} 1, & s \in E_- \\ -i, & s \in F_- \\ -1, & s \in E_+ \\ -i, & s \in F_+ \end{cases}$$

So

$$(kh\widehat{\eta}_{\gamma})(s) = \begin{cases} \frac{1}{\sqrt{2\pi}}\cos\gamma(s), & s \in E_- \\ \frac{1}{\sqrt{2\pi}}\sin\gamma(2s), & s \in F_- \\ -\frac{1}{\sqrt{2\pi}}\cos\gamma(2s-4\pi) & s \in E_+ \\ \frac{1}{\sqrt{2\pi}}\sin\gamma(s-4\pi), & s \in F_+ \\ 0, & \text{otherwise.} \end{cases}$$

This shows that Fourier transforms of wavelets in our class can be *real-valued*. For this to be continuous, computations at $s = -\frac{8\pi}{3}, -\frac{4\pi}{3}, \frac{4\pi}{3}$ would imply

$$\cos\gamma(-\frac{8\pi}{3}) = 0, \cos\gamma(-\frac{4\pi}{3}) = \sin\gamma(-\frac{8\pi}{3}),$$

and

$$-\cos\gamma(-\frac{4\pi}{3}) = \sin\gamma(-\frac{8\pi}{3}),$$

an impossibility. So no (Fourier transform of a) wavelet $hk\widehat{\eta}_{\gamma}$ can be continuous.

Next, define a wavelet multiplier q by setting $q(s) := e^{\frac{is}{2}}$ for $s \in F$ and extending 2π-periodically to $\mathbb{R}$. Then on E_+ we have $q(s) = e^{\frac{i(s-2\pi)}{2}} = -e^{\frac{is}{2}}$, and on E_- we have $q(s) = e^{\frac{i(s+4\pi)}{2}} = e^{\frac{is}{2}}$. Then the (Fourier transform of a) wavelet

$$\widehat{\xi}_\gamma(s) := (qhk\widehat{\eta}_\gamma)(s) = \begin{cases} \frac{1}{\sqrt{2\pi}} e^{\frac{is}{2}} \cos\gamma(s), & s \in E_- \\ \frac{1}{\sqrt{2\pi}} e^{\frac{is}{2}} \sin\gamma(2s), & s \in F_- \\ \frac{1}{\sqrt{2\pi}} e^{\frac{is}{2}} \cos\gamma(2s - 4\pi), & s \in E_+ \\ \frac{1}{\sqrt{2\pi}} e^{\frac{is}{2}} \sin\gamma(s - 4\pi), & s \in F_+ \\ 0, & \text{otherwise.} \end{cases}$$

can be continuous. If γ is any continuous real-valued function on $[-\frac{8\pi}{3}, -\frac{4\pi}{3}]$ with $\gamma(-\frac{8\pi}{3}) = \frac{\pi}{2}$ and $\gamma(-\frac{4\pi}{3}) = 0$ then $\widehat{\xi}_\gamma$ is continuous on $\mathbb{R}$, and will be in class C^k if $\gamma(s) \in C^k$ and if the first k right/left derivatives of $\gamma(s)$ vanish at these two points. Wavelets of Meyer's class $\widehat{\psi}_{Me}$ are of this form. Indeed, if we let

$$\gamma(s) = \frac{\pi}{2}\nu(-\frac{3s}{4\pi} - 1),$$

then

$$\cos\gamma(2s - 4\pi) = \sin[\frac{\pi}{2}\nu(\frac{3s}{2\pi} - 1)] \quad \text{and} \quad \sin\gamma(s - 4\pi) = \cos[\frac{\pi}{2}\nu(\frac{3s}{4\pi} - 1)],$$

where we use the property $\nu(s) + \nu(1 - s) = 1$, hence

$$\widehat{\xi}_\gamma = \widehat{\psi}_{Me}.$$

Example 5.14. Meyer's class ψ_{Me} effectively demonstrates Corollary 5.12. We have

$$\text{supp}(\psi_{Me}) = E \cup F = [-\frac{8\pi}{3}, -\frac{2\pi}{3}) \cup [\frac{2\pi}{3}, \frac{8\pi}{3})$$

with E, F as in Example 5.13. We have

$$\Lambda(E, F) = [-\frac{8\pi}{3}, -\frac{4\pi}{3}).$$

If h is an arbitrary bounded function on $[-\frac{8\pi}{3}, -\frac{4\pi}{3})$, note that $[-\frac{8\pi}{3}, -\frac{4\pi}{3}) + 4\pi = [\frac{4\pi}{3}, \frac{8\pi}{3}) \subset E \cup F$, and extend h by defining $h(s) = h(s - 4\pi)$ for $s \in [\frac{4\pi}{3}, \frac{8\pi}{3})$. Then note that $K := [-\frac{8\pi}{3}, -\frac{4\pi}{3}) \cup [\frac{4\pi}{3}, \frac{8\pi}{3})$ is a 2-dilation generator of a partition of $\mathbb{R}\backslash\{0\}$, and extend h to $\mathbb{R}$ by $h(s) = h(2^{-n}s)$, $s \in 2^n K$, $n \in \mathbb{Z}$ and $h(0) = 0$. Then h is 2-dilation-periodic and if $s \in [-\frac{4\pi}{3}, -\frac{2\pi}{3})$, then $s + 2\pi \in [\frac{2\pi}{3}, \frac{4\pi}{3})$, hence $2s + 4\pi \in [\frac{4\pi}{3}, \frac{8\pi}{3})$. So $h(s + 2\pi) = h(2s + 4\pi) = h(2s) = h(s)$ as required. The reasoning is similar for $s \in [\frac{2\pi}{3}, \frac{4\pi}{3})$. So if $k(s)$ is defined on $[-\frac{8\pi}{3}, -\frac{2\pi}{3})$, a 2π-translation generator of a partition of $\mathbb{R}$, by $k(s) = h(s)$, and extended 2π-periodically to $\mathbb{R}$, then $k(s) = h(s)$ also on $[\frac{2\pi}{3}, \frac{8\pi}{3})$. That is, h agrees with k on $E \cup F$. Thus by Corollary 3.15, $M_h \in \{\widehat{D}, \widehat{T}, \widehat{P}_\psi\}'$. Moreover, by the uniqueness part of this construction, every operator in $\{\widehat{D}, \widehat{T}, \widehat{P}_\psi\}'$ has this form.

The following is simple but useful.

Proposition 5.15. *Let (E, F) be an interpolation pair of wavelet sets. Suppose h_1, h_2 are as in $(**)$. Then the wavelet $\widehat{\psi}$ in $(*)$ satisfies (a.e.)*

$$|\widehat{\psi}(s)|^2 + |\widehat{\psi}(\sigma_E^F(s))|^2 = \begin{cases} \frac{1}{2\pi}, & s \in E \setminus F \\ \frac{1}{2\pi}, & s \in F \setminus E \\ \frac{1}{\pi}, & s \in E \cap F \\ 0, & otherwise \end{cases}$$

[Note: this does *not* imply that $\widehat{\psi}$ is discontinuous, since the discontinuity of $|\widehat{\psi}|^2 + |\widehat{\psi} \circ \sigma_E^F|^2$ can be due entirely to the discontinuity of σ_E^F. This is the case for ψ_{Me} in particular.]

Proof. We have

$$\widehat{\psi}(s) = \begin{cases} \frac{1}{\sqrt{2\pi}} h_1(s), & s \in E \setminus F \\ \frac{1}{\sqrt{2\pi}} h_2(s), & s \in F \setminus E \\ \frac{1}{\sqrt{2\pi}}(h_1(s) + h_2(s)), & s \in E \cap F. \end{cases}$$

If $s \in E \setminus F$ then $\sigma_E^F(s) \in F \setminus E$. So

$$\widehat{\psi}(\sigma_E^F(s)) = \frac{1}{\sqrt{2\pi}} h_2(\sigma_E^F(s)).$$

Since

$$\begin{pmatrix} h_1 & h_2 \\ h_2 \circ \sigma_E^F & h_1 \circ \sigma_E^F \end{pmatrix}$$

is a unitary matrix a.e., we must have

$$|h_1(s)|^2 + |h_2(\sigma_E^F(s))|^2 = 1 \text{ a.e.}$$

Thus

$$|\widehat{\psi}(s)|^2 + |\widehat{\psi}(\sigma_E^F(s))|^2 = \frac{1}{\sqrt{2\pi}} \text{ a.e.}$$

Similarly, if $x \in F \setminus E$ the same equality holds. If $s \in E \cap F$, then $\sigma_E^F(s) = s$. The matrix $(**)$ is then

$$\begin{pmatrix} h_1(s) & h_2(s) \\ h_2(s) & h_1(s) \end{pmatrix}.$$

The condition that this is unitary implies that $|h_1(s) + h_2(s)| = 1$, hence

$$|\widehat{\psi}(s)|^2 = \frac{1}{2\pi} = |\widehat{\psi}(\sigma_E^F(s))|^2. \quad \square$$

Let $\partial(K) = \overline{K} \cap \overline{K^c}$ denote the boundary of a set K.

Corollary 5.16. *Let (E, F) be an interpolation pair. Suppose E, F are finite unions of intervals. A necessary condition for there to exist h_1, h_2 satisfying $(**)$ such that $\widehat{\psi}$ of $(*)$ is continuous on $\mathbb{R}$ is that*

$$\partial(E \cup F) \cap \sigma_E^F(\partial(E \cup F)) = \emptyset.$$

Proof. If $\widehat{\psi}$ is continuous then $\widehat{\psi}$ must vanish on $\partial(E \cup F)$. If $s \in E \cup F$ then Proposition 5.15 shows that $\widehat{\psi}$ cannot vanish at both s and $\sigma_E^F(s)$. So either

$$s \notin \partial(E \cup F)$$

or
$$\sigma_E^F(s) \notin \partial(E \cup F).$$

Now suppose $s_0 \in \partial(E \cup F)$ is arbitrary. Since E and F are wavelet sets which are finite unions of intervals, s_0 must be a limit of points $s_n \in E \cup F$ on which $(\sigma_E^F(s) - s)$ is constant. Then

$$\sigma_E^F(s_n) \to \sigma_E^F(s_0),$$

so

$$\widehat{\psi}(\sigma_E^F(s_n)) \to \widehat{\psi}(\sigma_E^F(s_0)).$$

Also

$$\widehat{\psi}(s_n) \to \widehat{\psi}(s_0).$$

Thus

$$|\widehat{\psi}(\sigma_E^F(s_n))|^2 + |\widehat{\psi}(s_n)|^2 \to |\widehat{\psi}(\sigma_E^F(s_0))|^2 + |\widehat{\psi}(s_0)|^2.$$

By Proposition 5.15,

$$|\widehat{\psi}(\sigma_E^F(s_n))|^2 + |\widehat{\psi}(s_n)|^2 \geq \frac{1}{2\pi}$$

for all n. Since $\widehat{\psi}(s_0) = 0$, this implies that $\widehat{\psi}(\sigma_E^F(s_0)) \neq 0$. Thus

$$\sigma_E^F(s_0) \notin \partial(E \cup F). \quad \square$$

The above Proposition raises some questions: (1) Is the necessary condition in Corollary 5.16 sufficient? (2) Can the hypothesis that E, F are finite unions of intervals be removed?

Example 5.17. The class of interpolation pairs given in A.1 of the Appendix gives a good demonstration of Corollary 5.16 and yields a natural generalization of the class ψ_{Me}. For $0 < \alpha < \frac{\pi}{3}$, follow the scheme of Example 5.13, letting

$$E = E_{-\alpha} = [-2\pi - 2\alpha, -\pi - \alpha) \cup [\pi - \alpha, 2\pi - 2\alpha)$$

and

$$F = E_\alpha = [-2\pi + 2\alpha, -\pi + \alpha) \cup [\pi + \alpha, 2\pi + 2\alpha).$$

The description of $\widehat{\eta}_\gamma(s)$ is analogous, with the exception that on $E \cap F = [-2\pi + 2\alpha, -\pi - \alpha) \cup [\pi + \alpha, 2\pi - 2\alpha)$, $\widehat{\eta}_\gamma(s) = \cos\gamma(s) + i\sin\gamma(s)$. Then $|\widehat{\eta}_\gamma(s)| = 1$ on $E \cap F$. Define $k(s)$ analogously, with the exception that on $E \cap F$ let $k(s) = e^{\frac{is}{2}}(\widehat{\eta}_\gamma(s))^{-1}$. Let $h(s)$ and $q(s)$ be as in Example 5.13. Then $\widehat{\xi}_\gamma := qhk\widehat{\eta}_\gamma$ has the form

$$\widehat{\xi}_\gamma(s) = \begin{cases} \frac{1}{\sqrt{2\pi}}e^{\frac{is}{2}}\cos\gamma(s), & s \in [-2\pi - 2\alpha, -2\pi + 2\alpha) \\ \frac{1}{\sqrt{2\pi}}e^{\frac{is}{2}}, & s \in [-2\pi + 2\alpha, -\pi - \alpha) \\ \frac{1}{\sqrt{2\pi}}e^{\frac{is}{2}}\sin\gamma(2s), & s \in [-\pi - \alpha, -\pi + \alpha) \\ \frac{1}{\sqrt{2\pi}}e^{\frac{is}{2}}\cos\gamma(2s - 4\pi), & s \in [\pi - \alpha, \pi + \alpha) \\ \frac{1}{\sqrt{2\pi}}e^{\frac{is}{2}}, & s \in [\pi + \alpha, 2\pi - 2\alpha) \\ \frac{1}{\sqrt{2\pi}}e^{\frac{is}{2}}\sin\gamma(s - 4\pi), & s \in [2\pi - 2\alpha, 2\pi + 2\alpha) \\ 0, & \text{otherwise.} \end{cases}$$

As in Example 5.13, if γ is any continuous real-valued function on $[-2\pi - 2\alpha, -2\pi + 2\alpha)$ with $\gamma(-2\pi - 2\alpha) = \frac{\pi}{2}$ and $\gamma(-2\pi + 2\alpha) = 0$, it is easily checked that $\widehat{\xi}_\gamma(s)$ is continuous on $\mathbb{R}$. For small $\alpha > 0$, $\widehat{\xi}_\gamma(s)$ can be thought of as a wavelet obtained from the Littlewood-Paley wavelet multiplied by $e^{\frac{is}{2}}$ (a wavelet by Remark 4.4) by "rounding down the corners" appropriately. Also note that by modifying $k(s)$ we can replace $\frac{1}{\sqrt{2\pi}}e^{is}$ on $E \cap F$ in the description of $\widehat{\xi}_\gamma(s)$ by any function $\delta(s)$ of uniform modulus $\frac{1}{\sqrt{2\pi}}$ that is continuous on the closure of $E \cap F$ and which takes the same values at the endpoints, and still achieve continuity of $\widehat{\xi}_\gamma$. We can achieve regularity properties by appropriate choices of the parameters, as with $\widehat{\psi}_{Me}$.

We conclude with a problem.

Problem F. *If (E, F) is an interpolation pair of wavelet sets, suppose f, g are functions in $L^\infty(\mathbb{R})$ such that*

$$f\widehat{\psi}_E + g\widehat{\psi}_F$$

is the Fourier transform of a wavelet. Are there 2-dilation-periodic functions $h_1, h_2 \in L^\infty(\mathbb{R})$ with

$$\widehat{\psi} = h_1\widehat{\psi}_E + h_2\widehat{\psi}_F$$

such that $\{h_1, h_2\}$ satisfy the Coefficient Criterion $(**)$?

CHAPTER 6

Concluding Remarks

1. Unitary Equivalence

Unitary representations π_1, π_2 of a group G on Hilbert spaces $\mathcal{H}_1$, $\mathcal{H}_2$ are called *equivalent* if there is a unitary operator $W \in \mathcal{B}(\mathcal{H}_1, \mathcal{H}_2)$ such that

$$W\pi_1(g)x = \pi_2(g)Wx, \quad g \in G, \ x \in \mathcal{H}_1.$$

Let G_0 be an allowable subset of G in the sense of the remark after Example 1.11, and suppose π_1, π_2 are wandering vector representations of (G, G_0) with complete wandering vectors ψ_1, ψ_2, respectively for $\pi_1(G_0)$, $\pi_2(G_0)$. There is a natural equivalence relation on pairs (π, ψ) which extends the usual equivalence relation for representations, namely, require that the unitary W implementing the equivalence of π_1 and π_2 satisfy $W\psi_1 = \psi_2$. If π is fixed, then the corresponding induced equivalence relation on $\mathcal{W}(\pi(G_0))$ is that $\psi_1 \sim \psi_2$ iff there exists $V \in \mathbb{U}(\pi(G_0)')$ with $V\psi_1 = \psi_2$. (This is analogous to the equivalence of MRA's introduced in [15], with scaling functions ϕ_i in place of the ψ_i.) This is equivalent to the condition that the vector states ω_{ψ_1} and ω_{ψ_2} agree on $C^*(\pi(\mathcal{G}))$. If $\mathbb{U}(\mathcal{C}_{\psi_1}(\pi(G_0))) \neq \mathbb{U}(\pi(G_0)')$, as is most often the case by Proposition 1.7, then there exist *inequivalent* complete wandering vectors, as happens for $\langle D, T \rangle$. An application of Proposition 1.3 then shows that the set of equivalence classes of $\mathcal{W}(\pi(G_0))$ can be parametrized by the coset space

$$\mathbb{U}(\mathcal{C}_\psi(\pi(G_0)))/\mathbb{U}(\pi(G_0)')$$

for any single choice of $\psi \in \mathcal{W}(\pi(G_0))$. This raises a natural *Question:* If $\mathcal{U} \subseteq \mathcal{B}(\mathcal{H})$ is a unitary system, when is $\mathbb{U}(\mathcal{U}')$ normalized by every element of $\mathbb{U}(\mathcal{C}_\psi(\mathcal{U}))$? Equivalently, when is $\mathbb{U}(\mathcal{U}')$ a normal subgroup of $\mathrm{Group}(\mathbb{U}(\mathcal{C}_\psi(\mathcal{U})))$? Is this true for the wavelet case $\mathcal{U} = \mathcal{U}_{D,T}$? Theorem 5.3 sheds some light in that it shows that if $\psi = \psi_E$ is s-elementary, then the *interpolation* unitaries in $\mathcal{C}_\psi(D, T)$ all normalize $\mathcal{U}' = \{D, T\}'$.

In a related direction, if ψ and η are inequivalent elements of $\mathcal{W}(\mathcal{U})$, then there is an element $V \in \mathrm{Group}(\mathcal{U})$ such that $\langle V\psi, \psi \rangle \neq \langle V\eta, \eta \rangle$. So the equivalence classes of $\mathcal{W}(\mathcal{U})$ are completely parameterized by the vector functions

$$\omega_\psi : \mathrm{Group}(\mathcal{U}) \to \mathbb{C}$$

given by $\omega_\psi(V) = \langle V\psi, \psi \rangle$. As noted in the introduction, we have

$$\mathrm{Group}(\mathcal{U}_{D,T}) = \{D^n T_\beta : n \in \mathbb{Z}, \beta \in \mathbb{D}\}.$$

So the function $\lambda_\psi : \mathbb{Z} \times \mathbb{D} \to \mathbb{C}$ given by $\lambda_\psi(n, \beta) = \langle D^n T_\beta \psi, \psi \rangle$ is a complete unitary equivalence invariant for wavelets. Two rather immediate properties are: (1) $|\widehat{\psi}| = |\widehat{\eta}|$ if and only if $\lambda_\psi(0, \beta) = \lambda_\eta(0, \beta)$, $\beta \in \mathbb{D}$, and (2) $|\widehat{\psi}| = \frac{1}{\sqrt{2\pi}}\chi_E$ for some wavelet set E if and only if $\lambda_\psi(n, \beta) = 0$ whenever $n \neq 0$. It may be worthwhile to explore this invariant further.

53

2. Higher Dimensional Systems

Let m be a positive integer, and let T_k be the unitary operator of translation by 1 in the kth coordinate direction on $L^2(\mathbb{R}^m)$. Let A be an $m \times m$ invertible matrix with real entries. Then the operator $D : L^2(\mathbb{R}^m) \to L^2(\mathbb{R}^m)$ defined by

$$(Df)(x) = |det A|^{\frac{1}{2}} f(Ax), \ x \in \mathbb{R}^m, f \in L^2(\mathbb{R}^m),$$

is easily shown to be unitary. If A is a *strict dilation* in the sense that A^{-1} is a strict contraction (i.e. $\|A^{-1}\| < 1$), or more generally if the eigenvalues of A all have modulii strictly greater than 1 (so D is *similar* to a strict dilation), then A *dilates* the unit ball B_1 of $\mathbb{R}^m$ in the sense that

$$\cup_{n=1}^{\infty} A^n B_1 = \mathbb{R}^m.$$

In this case it can be shown that the unitary system

$$\mathcal{U}_{D,T_1,T_2,\cdots,T_m} := \{D^n T_1^{l_1} T_2^{l_2} \cdots T_m^{l_m} : n, l_i \in \mathbb{Z}, 1 \le i \le m\}$$

has complete wandering vectors. There are *also* called orthogonal wavelets, as in the 1-dimensional case. (For all choices of A orthogonal wavelets exist. For some choices of A, but not for all, it can be shown that such wavelets exist which have "good" regularity properties, so are useful in applications.) Using the n-dimensional Fourier transform (which is the tensor product of n copies of the 1-dimensional Fourier transform) essentially all of Chapter 3 generalizes to this setting. Indeed, since the unitaries T_i commute the unitary system $\mathcal{U}_{D,T_1,T_2,\cdots,T_m}$ factors as the product of two abelian groups, and so the structural results of Chapter 2 all apply, yielding analogs of Lemma 3.1 and Theorem 3.9, exactly as in Chapter 3. We leave proofs of the details and extensions of the other results, to the reader. The definition of *wavelet set,* and of s-elementary wavelet, makes sense for these dilation-translation systems in $\mathbb{R}^m$, and it can be shown that they always exist. (A proof of existence of wavelet sets for general A has been obtained by the present authors together with D. Speegle, who is at present a graduate student at Texas A&M University. In addition, Speegle has obtained a proof that the family of s-elementary wavelets is a connected subset of the unit ball of $L^2(\mathbb{R}^m)$.) Much of the interpolation theory of Chapter 5 (formally) adapts to this setting, and we leave details to the reader.

3. Multiresolution Analysis

The method of multiresolution analysis is a very important technique for deriving wavelets. In fact, in many respects the formulation of MRA (c.f. [**8, 19, 21**]) captures the "essence" of wavelet theory. The MRA technique is a method of starting with a suitably chosen unit vector ϕ for which the set $\{T^l \phi : l \in \mathbb{Z}\}$ is a Riesz basis for its closed span in $L^2(\mathbb{R})$, and using it to derive a complete wandering vector ψ for $\mathcal{U}_{D,T}$. Suitably "chosen" means that the 2-scale relation $D^{-1}\phi \in \overline{\mathrm{span}}\{T^l \phi : l \in \mathbb{Z}\}$ is satisfied and that $\overline{\mathrm{span}}\{D^n T^l \phi : n \in \mathbb{Z}_+, l \in \mathbb{Z}\} = L^2(\mathbb{R})$, where $\mathbb{Z}_+ = \{0, 1, 2, \cdots\}$. This method can be given an operator-theoretic formulation. (c.f. [**6, 15**]). However, its connection with the local commutant, which is the main new tool used in this paper, is at most indirect. The reason for this is that since $\{D^n T^l : n \in \mathbb{Z}_+, l \in \mathbb{Z}\}$ is a semigroup (apply Lemma 3.2) we have $\mathcal{C}_\phi(\mathcal{U}_{D,T}) = \{D, T\}'$ (see the remark after Lemma 1.1), so two inequivalent scaling functions cannot be related via a unitary in the local commutant of $\{D, T\}$ at one of them.

4. A Connection With Some Work of Guido Weiss

Several months after this manuscript was submitted we discovered that there is a connection between some of the work we have presented in this article and some of the work of Guido Weiss and his group who have been working on a program of a unified approach to wavelet theory via the Fourier transform. We learned of this connection in late spring 1995, when our colleagues Charles Chui, Bill Johnson, and John Mc Carthy saw talks each of us gave separately, and alerted both Larson and Weiss. We thank our colleagues very much for pointing this out. In the past year we have had some fruitful interaction between our respective groups concerning this.

It turned out that the class we call s-elementary wavelets in Chapter 4 were also discovered completely independently, and in about the same time period, by Professor Weiss, his colleague E. Hernandez (U. Madrid), and Weiss' students X. Fang and X. Wang. They called them MSF wavelets, and they were introduced in a series of three papers, the first of which is due to Fang and Wang (Construction of minimally supported frequency wavelets, J. Fourier Anal. Appl. 2 (1996) 315-327) and the second and third to Hernandez, Wang and Weiss (Smoothing minimally supported frequency (MSF) wavelets: Parts I and II, J. Fourier Anal. Appl. (1996)). The first of these was submitted within two days of submission of this memoir. As we noted in Remark 4.4 , any function supported on a wavelet set which is unimodular on the wavelet set is the Fourier transform of a wavelet. The class of MSF wavelets were defined to include this more general type also. So our wavelet sets are the just the support sets of Fourier transforms of MSF wavelets.

The reasons that the two groups were led to consider this special class of wavelets were different. In Professor Weiss' program techniques were developed to "smooth" appropriate MSF wavelets to produce new wavelets with continuous Fourier transform, and this led to new results concerning multiresolution analysis (MRA) wavelets. In our case, we *greatly* needed concrete examples of pairs of wavelets with simple enough structure to enable us to experiment by hand with operator computations, and we found that we could modify the Littlewood-Paley set.

However, there is some definite overlap in the results we obtained. The characterization in this paper (Lemma 4.3) of wavelet sets as measurable sets which generate measurable partitions of the line under both 2-dilation and 2π-translation is also a result of Fang and Wang. They also obtained an example of an unbounded wavelet set. It is different from Example 4.5(xi) in this paper. There are some overlaps in the classes of examples of wavelet sets constructed, but there are also big differences. The smoothing technique of Weiss and our method of interpolation have apparently very little in common, yet they lead to some of the same classes of wavelets. Both give new ways of obtaining Meyer's family: ours by interpolating between two wavelet sets and their's by smoothing Shannon's wavelet. At the same time, each seems to yield classes that apparently cannot be obtained using the others' techniques. Most notably, we seem to have been giving our research students some of the same wavelet problems to work on. This fact became apparent during a conference last summer (see 6.7.3). We note that the connectedness problem (i.e. Problem A in this paper) was also raised by Professor Weiss.

We remark that much of the above mentioned work of Professor Weiss and his group (but not the overlap with ours) are detailed in the recent excellent book (E. Hernandez and G. Weiss, A First Course on Wavelets, CRC Press, Boca Raton). Professor Weiss mentioned to us last summer that, because of the short timing involved in our realization of the extent of overlap between our groups, it was not possible to insert appropriate comments concerning this in their book before it appeared. He asked us to instead insert a description of this matter in our memoir before publication. We thank him for his support of our work.

5. Status of Problems

In this manuscript we had posed a number of problems, with the main ones designated by the letters A - F. In the past year, problems B, C, D, the finite group case of E, and F have been solved, although they seemed hopelessly out of reach at the time this paper was written. This reveals something about how rapidly the theory is evolving. Problem A is the connectedness problem, and at the time of this writing it is still open, although much positive progress has been made. This problem was also raised independently by Professor Weiss. D. Han and V. Kamat in their thesis work at Texas A&M University, and independently W.S. Li, J. Mc Carthy and D. Timotin, have proven that Theorem 2.16 is valid when $\mathcal{U}_0$ is nonabelian. Both groups used this result to anwer Problem B negatively. Problem C was answered negatively by D. Han, and independently by Li, McCarthy and Timotin; the counterexamples are different. Problem C' and D were answered negatively by Dai, Gu, Larson and Liang. The finite group case of Problem E was answered positively by Q. Gu in his thesis work at A&M; he showed that every finite group is attainable. Problem F was answered positively by Gu and Larson.

6. Examples

The referee of this manuscript kindly suggested that we include some additional material, and in particular it was strongly suggested that we include some concrete examples of wavelet sets in the plane. The existance proof, mentioned above, is contained in the preprint (*"Wavelet sets in $\mathbb{R}^n$ "*, by X. Dai, D. Larson and D. Speegle, which is to appear in J. Fourier Anal. Appl.). It is basically constructive, but the constructive technique does not directly yield any examples of "elegance". So we follow the referee's advice and include two examples that we worked out (except for the graph) shortly after this manuscript was accepted for publication in October 1995. We remark that, motivated by our existence result, other examples have been worked out by P. Soardi and D. Weiland, and Q. Gu and D. Speegle, and perhaps others. In fact, the graph we inserted in Example 6.6.1 (but not the example itself) was inspired by the beautiful graphs in the recent preprint by Soardi and Weiland entitled *"Single wavelets in n-dimensions"* in which precise methods are detailed for constructing a family of MSF wavelets (and hence wavelet sets) with fractal-like nature in the plane and in higher dimensions for dilation factor $2I$. In addition, we note that Gu and Speegle have shown (not yet published) that interpolation pairs of wavelet sets exist in the plane for certain matricial dilation factors, and hence that single-function wavelets exist in $L^2(\mathbb{R}^2)$ which are not MSF wavelets.

Example 6.6.1 Consider the dilation matrix $A = 2I$, where I is the identity matrix on $\mathbb{R}^2$. For $n \in \mathbb{N}$, define 2-dimensional vectors $\vec{\alpha}_n, \vec{\beta}_n$ by:

$$\vec{\alpha}_n \quad := \quad \frac{1}{4^{n-1}}\left(\frac{\pi}{2}, \frac{\pi}{2}\right) \in \mathbb{R}^2$$

$$\vec{\beta}_n \quad := \quad \sum_{k=1}^{n} \vec{\alpha}_k,$$

$$\vec{\beta}_0 \quad := \quad 0.$$

Define

$$G_0 \quad = \quad [0, \frac{\pi}{2}] \times [0, \frac{\pi}{2}]$$

$$G_n \quad = \quad \frac{1}{4^n}G_0 + \vec{\beta}_n;$$

$$E \quad = \quad \bigcup_{k=1}^{\infty} G_k \subset 2G_0 \setminus G_0;$$

$$C \quad = \quad G_0 \cup E + (2\pi, 2\pi);$$

$$B \quad = \quad 2G_0 \setminus (G_0 \cup E).$$

Define

$$A_1 \quad = \quad B \cup C$$

$$A_2 \quad = \quad \{(-x, y) : (x, y) \in A_1\}$$

$$A_3 \quad = \quad \{(-x, -y) : (x, y) \in A_1\}$$

$$A_4 \quad = \quad \{(x, -y) : (x, y) \in A_1\}$$

$$W_1 \quad = \quad A_1 \cup A_2 \cup A_3 \cup A_4.$$

Then the set W_1 is a 2-dimensional wavelet set for the system $\mathcal{U}_{D,T_1,T_2}$ in section 6.2, where $(Df)(x) = |\det A|^{1/2} f(Ax) = 2f(2x), x \in \mathbb{R}^2, f \in L^2(\mathbb{R}^2)$.

Proof. We have

$$4E \quad = \quad (4G_1) \cup \left(\bigcup_{k=2}^{\infty} 4G_k\right)$$

$$= \quad (G_0 + (2\pi, 2\pi)) \cup \left(\bigcup_{k=2}^{\infty} \frac{1}{4^{k-1}}G_0 + \vec{\beta}_{k-1} + (2\pi, 2\pi)\right)$$

$$= \quad (2\pi, 2\pi) + G_0 \cup E$$

$$= \quad C.$$

Since

$$E \cup B = 2G_0 \setminus G_0$$

is a 2-dilation generator for the first quadrant, and $4E = C$, so A_1 is a 2-dilation generator for this. So W_1 is a 2-dilation generator for the 2-dimensional plane $\mathbb{R}^2$.

Also, it is clear that W_1 is congruent modulo translations along the coordinate axes by integral multiples of 2π to the set $[-\pi, \pi] \times [-\pi, \pi]$. $\square$

The following graph was inspired by the graphs of Soardi and Weiland, as noted above.

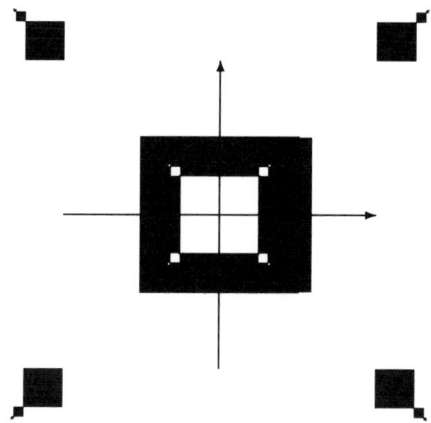

Figure 1: Wavelet set W_1 in $\mathbb{R}^2$

Example 6.6.2 Let A and E be as in Example 6.6.1. For $n \in \mathbb{N}$, define 2-dimensional vectors $\vec{\alpha}_n, \vec{\beta}_n$ by:

$$\vec{\alpha}_n := \frac{1}{4^{n-1}} (\frac{\pi}{2}, 0) \in \mathbb{R}^2,$$

$$\vec{\beta}_n := \sum_{k=1}^{n} \vec{\alpha}_k,$$

$$\vec{\beta}_0 := 0.$$

Define

$$G_0 = [0, \frac{\pi}{2}] \times [-\frac{\pi}{2}, \frac{\pi}{2}]$$

$$G_n = \frac{1}{4^n} G_0 + \vec{\beta}_n;$$

$$E = \bigcup_{k=1}^{\infty} G_k \subset 2G_0 \backslash G_0;$$

$$C = G_0 \cup E + (2\pi, 0);$$

$$B = 2G_0 \backslash (G_0 \cup E).$$

Define

$$A_1 = B \cup C$$

$$A_2 = \{(-x, y) : (x, y) \in A_1\}$$

$$W_2 = A_1 \cup A_2.$$

Then the set W_2 is a 2-dimensional wavelet set.

Proof. We have

$$
\begin{aligned}
4E &= (4G_1) \cup (\bigcup_{k=2}^{\infty} 4G_k) \\
&= (G_0 + (2\pi, 0)) \cup (\bigcup_{k=2}^{\infty} \frac{1}{4^{k-1}} G_0 + \vec{\beta}_{k-1} + (2\pi, 0)) \\
&= (2\pi, 0) + G_0 \cup E \\
&= C.
\end{aligned}
$$

Since

$$
E \cup B = 2G_0 \backslash G_0
$$

is a 2-dilation generator for the right half plane, and $4E = C$, so A_1 is a 2-dilation generator for this. So W_2 is a 2-dilation generator for the 2-dimensional plane $\mathbb{R}^2$.

Also, it is clear that W_2 is congruent modulo translations along the coordinate axes by integral multiples of 2π to the set $[-\pi, \pi] \times [-\pi, \pi]$. $\square$

7. Acknowledgments

7.1. We wish to thank the referee for some excellent suggestions for additional further directions of research that we were either unaware of or only partially aware of. We take the opportunity to note that some of these ideas have significantly influenced the subsequent projects we have undertaken.

7.2. The work in this manuscript was begun in 1992 when the first author was a summer participant in the NSF [MSRG] funded Workshop in Linear Analysis and Probability at Texas A&M University.

7.3. The work in this manuscript was presented, in various stages of development, in a number of conference talks including AMS Special Sessions in October 1992 (Dayton), January 1993 (San Antonio), October 1993 (College Station), January 1994 (Cincinnati), and hour talks at GPOTS-94 (U. Nebraska) and SEAM-95 (Georgia Tech). Two talks by the second author on this and subsequent work were given at a NATO Advanced Study Institute on "*Operator Algebras and Applications*", August 1996 (Samos, Greece). (An expository conference proceedings article is available entitled "*Von Neumann algebras and wavelets*".) Special credit is due to the AFOSR, NSA and UNCC for funding a conference in April, 1993 (UNC-Charlotte) which featured early stages of this work among work of others. Credit is also due to the AFOSR, NSF and UNCC for sponsoring a similar conference on operator theory and wavelet theory in July, 1996, in Charlotte which promoted interaction between our research group and the group of Professor Weiss.

Appendix: Examples of Interpolation Maps

The class of interpolation maps between wavelet sets exhibits many different types of algebraic structural properties that are relevant to this article. We give some examples illustrating these properties. Many are referenced as counterexamples in this article. These examples also serve to illustrate the manner in which one can compute with interpolation maps.

Consider the path of wavelet sets of Example 4.5 (ii) given by

$$E_\alpha = [-2\pi + 2\alpha, -\pi + \alpha) \cup [\pi + \alpha, 2\pi + 2\alpha)$$

for $-\pi < \alpha < \pi$. The interpolation maps $\sigma_{\alpha_1}^{\alpha_2} := \sigma_{E_{\alpha_1}}^{E_{\alpha_2}}$ have some useful properties. First consider the special case where $\alpha_1 = -\alpha$ and $\alpha_2 = \alpha$. Here the union $E_1 \cup E_2$ is the symmetric set

$$[-2\pi - 2\alpha, -\pi + \alpha) \cup [\pi - \alpha, 2\pi + 2\alpha).$$

Example A.1. We will show that $\sigma_{-\alpha}^\alpha$ is involutive for α in the range $0 \le \alpha \le \frac{\pi}{3}$. Write $\sigma := \sigma_{-\alpha}^\alpha$. To compute σ we can partition

$$E_{-\alpha} = A_1 \cup B_1 \cup C_1 \cup D_1 \text{ and } E_\alpha = A_2 \cup B_2 \cup C_2 \cup D_2$$

where

$$A_1 = [-2\pi - 2\alpha, -2\pi + 2\alpha), \ B_1 = [-2\pi + 2\alpha, -\pi - \alpha),$$
$$C_1 = [\pi - \alpha, \pi + \alpha), \ D_1 = [\pi + \alpha, 2\pi - 2\alpha),$$

and

$$A_2 = [-2\pi + 2\alpha, -\pi - \alpha), \ B_2 = [-\pi - \alpha, -\pi + \alpha),$$
$$C_2 = [\pi + \alpha, 2\pi - 2\alpha), \ D_2 = [\pi + \alpha, 2\pi - 2\alpha).$$

Then

$$A_1 + 4\pi = D_2, \ B_1 = A_2, \ C_1 - 2\pi = B_2, \ D_1 = C_2,$$

and we also have the arithmetic relations $2B_2 = A_1$ and $D_2 = 2C_1$. Thus on A_1 we have $\sigma(s) = s + 4\pi \in D_2$, on C_1 we have $\sigma(s) = s - 2\pi \in B_2$, and $\sigma(s) = s$ for $s \in B_1 \cup D_1$. So for $s \in A_1$, since $\sigma(s) \in D_2$ and $D_2 = 2C_1$ we have $\frac{1}{2}\sigma(s) \in C_1$, so using the 2-homogeneity of σ we have

$$\sigma^2(s) = 2\sigma(\tfrac{1}{2}\sigma(s)) = 2(\tfrac{1}{2}\sigma(s) - 2\pi) = \sigma(s) - 4\pi = s + 4\pi - 4\pi = s.$$

Similarly, for $s \in C_1$, we have $2\sigma(s) \in A_1$, so

$$\sigma^2(s) = \tfrac{1}{2}\sigma(2\sigma(s)) = \tfrac{1}{2}(2\sigma(s) + 4\pi) = \sigma(s) + 2\pi = s - 2\pi + 2\pi = s.$$

On $B_1 \cup D_1$ we have $\sigma(s) = s$ so $\sigma^2(s) = s$. Thus $\sigma^2(s) = s$ for all $s \in E_{-\alpha}$, and hence $\sigma^2(s) = s$ for all $s \in \mathbb{R}$ since σ is determined by its restriction to $E_{-\alpha}$.

Example A.2. Next consider the example above, but for α in the range $\frac{\pi}{3} < \alpha \le \frac{\pi}{2}$. We will show that in this case σ is *not* involutive, and moreover, the restriction of σ^2 to $E_{-\alpha}$ is *not* a 2π-congruence. In addition the orbit of a point in $E_{-\alpha}$ can be unbounded. Here a different (from that of A.1) partitioning is required to compute $\sigma_{-\alpha}^{\alpha}$. Let

$$A_1 = [-2\pi - 2\alpha, -3\pi + \alpha), \ B_1 = [-3\pi + \alpha, -\pi - \alpha),$$
$$C_1 = [\pi - \alpha, 2\alpha), \ D_1 = [2\alpha, 2\pi - 2\alpha),$$

and

$$A_2 = [-2\pi + 2\alpha, -2\alpha), \ B_2 = [-2\alpha, -\pi + \alpha),$$
$$C_2 = [\pi + \alpha, 3\pi - \alpha), \ D_2 = [3\pi - \alpha, 2\pi + 2\alpha).$$

We have

$$A_1 + 2\pi = B_2, \ B_1 + 4\pi = C_2, \ C_1 + 2\pi = D_2, \ D_1 - 2\pi = A_2.$$

So

$$\sigma(s) = \begin{cases} s + 2\pi, & s \in A_1 \\ s + 4\pi, & s \in B_1 \\ s + 2\pi, & s \in C_1 \\ s - 2\pi, & s \in D_1. \end{cases}$$

The additional arithmetic relationships available in A.1 to compute σ^2 are not available here without further partitioning. The form of σ^2 is complicated. For instance, for $s = -2\pi - 2\alpha$, we have $\sigma(s) = s + 2\pi = -2\alpha$. Since $\frac{\pi}{3} \le \alpha < \frac{\pi}{2}$, we have $-4\alpha \in B_1$. So

$$\sigma^2(s) = \frac{1}{2}\sigma(-4\alpha) = -\frac{1}{2}(-4\alpha + 4\pi) = 2\pi - 2\alpha \in 2C_1.$$

So $\sigma^2(s) \ne s$ at this point (and at nearby points to the right of $s = -2\pi - 2\alpha$). Continuing, we have

$$\sigma^3(s) = \sigma(2\pi - 2\alpha) = 2\sigma(\pi - \alpha) = 2(\pi - \alpha + 2\pi) = 6\pi - 2\alpha.$$

Since $\frac{\pi}{3} < \alpha \le \frac{\pi}{2}$ we have $\pi - \alpha < \frac{(6\pi - 2\alpha)}{8} < 2\alpha$, so $\frac{1}{8}\sigma^3(s) \in C_1$, so

$$\sigma^4(s) = 8\sigma(\frac{1}{8}\sigma^3(s)) = 8(\frac{1}{8}\sigma^3(s) + 2\pi) = \sigma^3(s) + 16\pi = 22\pi - 2\alpha.$$

A simple induction argument shows that the orbit of $s = -2\pi - 2\alpha$ under σ is unbounded. Now, observe that since $\frac{\pi}{3} < \alpha \le \frac{\pi}{2}$ we have $-\frac{5\pi}{2} - \frac{\alpha}{2} \in A_1$. Let $s_1 = -\frac{5\pi}{2} - \frac{\alpha}{2}$. Then $\sigma(s_1) = s_1 + 2\pi = -\frac{\pi}{2} - \frac{\alpha}{2}$. So $4\sigma(s_1) = -2\pi - 2\alpha \in A_1$. So

$$\sigma^2(s_1) = \frac{1}{4}\sigma(4\sigma(s_1)) = \frac{1}{4}(4\sigma(s_1) + 2\pi) = \sigma(s_1) + \frac{\pi}{2} = s_1 + \frac{5\pi}{2}.$$

Thus $(\sigma^2(s_1) - s_1)/2\pi$ is not an integer. Hence σ^2 is *not* a 2π-congruence on $E_{-\alpha}$. So the corresponding composition operator U_σ is an element of the local commutant of $\{\widehat{D}, \widehat{T}\}$ at $\widehat{\psi}_{E_{-\alpha}}$ whose *square* is not in the local commutant.

Example A.3. Next, fix the initial wavelet set

$$E_0 = [-2\pi, -\pi) \cup [\pi, 2\pi),$$

and let the final wavelet set be E_α for $0 \le \alpha < \pi$. Write

$$\sigma_\alpha := \sigma_{E_0}^{E_\alpha}.$$

It is easy to compute that for $0 \leq \alpha \leq \frac{\pi}{2}$ we have, on E_0,

$$\sigma_\alpha(s) = \begin{cases} s + 4\pi, & s \in [-2\pi, -2\pi + 2\alpha) \\ s, & s \in [-2\pi + 2\alpha, -\pi) \\ s - 2\pi, & s \in [\pi, \pi + \alpha) \\ s, & s \in [\pi + \alpha, 2\pi) \end{cases}$$

and for $\frac{\pi}{2} < \alpha < \pi$ we have, on E_0,

$$\sigma_\alpha(s) = \begin{cases} s + 4\pi, & s \in [-2\pi, -\pi) \\ s + 2\pi, & s \in [\pi, 2\alpha) \\ s - 2\pi, & s \in [2\alpha, \pi + \alpha) \\ s, & s \in [\pi + \alpha, 2\pi) \end{cases}$$

For $0 \leq \alpha \leq \frac{\pi}{2}$ one may compute that σ_α is involutive. Indeed, if $s \in [\pi, \pi + \alpha)$ then

$$\sigma(s) = s - 2\pi \in [-\pi, -\pi + \alpha).$$

So $2\sigma(s) \in E_0$. We have

$$\sigma^2(s) = \frac{1}{2}\sigma(2\sigma(s)) = \frac{1}{2}(2\sigma(s) + 4\pi) = \sigma(s) + 2\pi = s.$$

The case $s \in [-2\pi, -2\pi + 2\alpha)$ is computed similarly. On the other hand, for $\frac{\pi}{2} < \alpha < \pi$ σ_α is not involutive. Indeed, for $s = \pi$ we have $\sigma(s) = s + 2\pi = 3\pi$, so $\frac{1}{2}\sigma(s) \in [2\pi, \pi + \alpha)$ and hence

$$\sigma^2(s) = 2\sigma(\frac{1}{2}\sigma(s)) = 2(\frac{3\pi}{2} - 2\pi) = -\pi.$$

Example A.4. For $0 \leq \alpha, \beta \leq \frac{\pi}{2}$ as above, one may compute that σ_α and σ_β *commute* as maps from $\mathbb{R}$ onto $\mathbb{R}$, so their corresponding composition operators $U_\alpha := U_{\sigma_\alpha}$ and $U_\beta := U_{\sigma_\beta}$ commute. Moreover, here the compositions $\sigma_\alpha \circ \sigma_\beta$ are all 2π-*congruences* of E_0, so

$$\{(\sigma_\alpha \circ \sigma_\beta)(E_0) : 0 \leq \alpha, \ \beta \leq \frac{\pi}{2}\}$$

is a 2-parameter family of wavelet sets. This is how the family $E_{\alpha\beta}$ if Example 4.5 (vi) was derived; Further compositions lead to multiparameter families. It follows that the group (under composition) generated by $\{\sigma_\alpha : 0 \leq \alpha \leq \frac{\pi}{2}\}$ is an abelian group of measure-preserving involutions of $\mathbb{R}$ such that for each element σ of the group the composition operator U_σ is contained in $\mathcal{C}_{\widehat{\psi}_0}(\widehat{D}, \widehat{T})$. Since each U_{σ_α} normalizes $\{\widehat{D}, \widehat{T}\}'$, so does U_σ. Letting

$$\mathcal{F} = \{\psi_\alpha : 0 \leq \alpha \leq \frac{\pi}{2}\},$$

this means that $\{\sigma_0, \mathcal{F}\}$ *admits interpolation*. It is, in a sense, prototypical of involutive interpolating families.

For $0 \leq \alpha < \beta \leq \frac{\pi}{2}$ one can compute, on E_0,

$$(\sigma_\alpha \circ \sigma_\beta)(s) = \begin{cases} s, & s \in [-2\pi, -2\pi + 2\alpha) \\ s + 4\pi, & s \in [-2\pi + 2\alpha, -2\pi + 2\beta) \\ s, & s \in [-2\pi + 2\beta, -\pi) \\ s, & s \in [\pi, \pi + \alpha) \\ s - 2\pi, & s \in [\pi + \alpha, \pi + \beta) \\ s, & s \in [\pi + \beta, 2\pi). \end{cases}$$

For instance, for $s \in [-2\pi + 2\alpha, -2\pi + 2\beta)$ we have $\sigma_\beta(s) = s + 4\pi \in [2\pi + 2\alpha, 2\pi + 2\beta)$, so $\frac{1}{2}\sigma_\beta(s) \in [\pi + \alpha, \pi + \beta) \subseteq [\pi + \alpha, 2\pi)$. So

$$(\sigma_\alpha \circ \sigma_\beta)(s) = 2\sigma_\alpha(\frac{1}{2}\sigma_\beta(s)) = 2(\frac{1}{2}\sigma_\beta(s)) = s + 4\pi.$$

Verifying commutativity on this interval, we have $\sigma_\alpha(s) = s$, so

$$(\sigma_\beta \circ \sigma_\alpha)(s) = \sigma_\beta(s) = s + 4\pi,$$

as required. The computations for the other subintervals of E_0 are similar.

Example A.5. For $0 \le \alpha$, $\beta < \pi$, if either α or β is greater then $\frac{\pi}{2}$ then σ_α and σ_β can fail to commute, and the composition can fail to act 2π-congruently on E_0. For instance, let $\alpha = \frac{\pi}{4}$ and $\beta = \frac{3\pi}{4}$. Then, on E_0,

$$\sigma_{\frac{\pi}{4}}(s) = \begin{cases} s + 4\pi, & s \in [-2\pi, -\frac{3\pi}{2}) \\ s, & s \in [-\frac{3\pi}{2}, -\pi) \\ s - 2\pi, & s \in [\pi, \frac{5\pi}{4}) \\ s, & s \in [\frac{5\pi}{4}, 2\pi) \end{cases}$$

and

$$\sigma_{\frac{3\pi}{4}}(s) = \begin{cases} s + 4\pi, & s \in [-2\pi, -\pi) \\ s + 2\pi, & s \in [\pi, \frac{3\pi}{2}) \\ s - 2\pi, & s \in [\frac{3\pi}{2}, \frac{7\pi}{4}) \\ s, & s \in [\frac{7\pi}{4}, 2\pi) \end{cases}$$

A routine computation yields

$$(\sigma_{\frac{\pi}{4}} \circ \sigma_{\frac{3\pi}{4}})(s) = \begin{cases} s, & s \in [-2\pi, -\frac{3\pi}{2}) \\ s + 4\pi, & s \in [-\frac{3\pi}{2}, -\pi) \\ s + 2\pi, & s \in [\pi, \frac{3\pi}{2}) \\ s - \pi, & s \in [\frac{3\pi}{2}, \frac{13\pi}{8}) \\ s - 2\pi, & s \in [\frac{13\pi}{8}, \frac{7\pi}{4}) \\ s, & s \in [\frac{7\pi}{4}, 2\pi). \end{cases}$$

and

$$(\sigma_{\frac{3\pi}{4}} \circ \sigma_{\frac{\pi}{4}})(s) = \begin{cases} s + 8\pi, & s \in [-2\pi, -\frac{3\pi}{2}) \\ s + 4\pi, & s \in [-\frac{3\pi}{2}, -\pi) \\ s, & s \in [\pi, \frac{5\pi}{4}) \\ s + 2\pi, & s \in [\frac{5\pi}{4}, \frac{3\pi}{2}) \\ s - 2\pi, & s \in [\frac{3\pi}{2}, \frac{7\pi}{4}) \\ s, & s \in [\frac{7\pi}{4}, 2\pi). \end{cases}$$

The map $\sigma = \sigma_{\frac{\pi}{4}} \circ \sigma_{\frac{3\pi}{4}}$ fails to act 2π-congruenctly on E_0 because for $s \in [\frac{3\pi}{2}, \frac{13\pi}{8})$,

$$\frac{(\sigma(s) - s)}{2\pi} = -\frac{1}{2} \notin \mathbb{Z}.$$

The image of E_0 under $\sigma_{\frac{\pi}{4}} \circ \sigma_{\frac{3\pi}{4}}$ is not a wavelet set because it is not 2π-translation congruent to $[0, 2\pi)$. So

$$U_{\sigma_{\frac{\pi}{4}}} U_{\sigma_{\frac{3\pi}{4}}} \notin \mathcal{C}_{\widehat{\psi}_0}(\widehat{D}, \widehat{T}).$$

On the other hand, $\sigma_{\frac{3\pi}{4}} \circ \sigma_{\frac{\pi}{4}}$ *is* a 2π-congruence of E_0. The image $(\sigma_{\frac{3\pi}{4}} \circ \sigma_{\frac{\pi}{4}})(E_0)$ is the wavelet set of Example 4.5 (v). We have

$$U_{\sigma_{\frac{3\pi}{4}}} U_{\sigma_{\frac{\pi}{4}}} \in \mathcal{C}_{\widehat{\psi}_0}(\widehat{D}, \widehat{T}).$$

This shows that the local commutant $\mathcal{C}_{\widehat{\sigma}_0}(\widehat{D}, \widehat{T})$ can contain unitaries A, B with $AB \in \mathcal{C}_{\widehat{\psi}_0}(\widehat{D}, \widehat{T})$ but

$$BA \notin \mathcal{C}_{\widehat{\psi}_0}(\widehat{D}, \widehat{T}).$$

In particular it shows that $\mathcal{C}_{\widehat{\psi}_0}(\widehat{D}, \widehat{T})$ is non-abelian.

Example A.6. Consider the map $\sigma_{\frac{3\pi}{4}}$ above. We can compute $\sigma_{\frac{3\pi}{4}}^2$, on E_0, as

$$\sigma_{\frac{3\pi}{4}}^2(s) = \begin{cases} s + 8\pi, & s \in [-2\pi, -\pi) \\ s - 2\pi, & s \in [\pi, \frac{3\pi}{2}) \\ s - \pi, & s \in [\frac{3\pi}{2}, \frac{7\pi}{4}) \\ s, & s \in [\frac{7\pi}{4}, 2\pi). \end{cases}$$

It follows that $\sigma_{\frac{3\pi}{4}}^3(s) = s$ for all $\in E_0$, and hence for all $s \in \mathbb{R}$. For instance, letting $\sigma := \sigma_{\frac{3\pi}{4}}$, for $s \in [-2\pi, -\pi)$ we have $\sigma^2(s) = s + 8\pi \in [6\pi, 7\pi)$. So $\frac{1}{4}\sigma^2(s) \in [\frac{3\pi}{2}, \frac{7\pi}{4}) \subset E_0$. Thus

$$\sigma^3(s) = 4\sigma(\frac{1}{4}\sigma^2(s)) = 4(\frac{1}{4}\sigma^2(s) - 2\pi) = \sigma^2(s) - 8\pi = s.$$

Computations for the other intervals are analogous. However, σ^2 is *not* a 2π-congruence of E_0 because $\sigma^2(s) - s$ is not an integral multiple of 2π for $s \in [\frac{3\pi}{2}, \frac{7\pi}{4})$. So $\sigma_{\frac{3\pi}{4}}$ does *not* admit operator-interpolation. We have $U_\sigma \in \mathcal{C}_{\widehat{\psi}_0}(\widehat{D}, \widehat{T})$ but

$$U_\sigma^* = U_\sigma^2 \notin \mathcal{C}_{\widehat{\psi}_0}(\widehat{D}, \widehat{T}),$$

showing that $\mathcal{C}_{\widehat{\psi}_0}(\widehat{D}, \widehat{T})$ is *not* self-adjoint.

Example A.7. Consider the set of Journe in Example 4.5 (i). Let

$$J = [-\frac{32\pi}{7}, -4\pi) \cup [-\pi, -\frac{4\pi}{7}) \cup [\frac{4\pi}{7}, \pi) \cup [4\pi, \frac{32\pi}{7}),$$

and let σ_J denote the interpolation map between E_0 and J. On E_0, we have

$$\sigma_J(s) = \begin{cases} s + 6\pi, & s \in [-2\pi, -\frac{10\pi}{7}) \\ s + 2\pi, & s \in [-\frac{10\pi}{7}, -\pi) \\ s - 2\pi, & s \in [\pi, \frac{10\pi}{7}) \\ s - 6\pi, & s \in [\frac{10\pi}{7}, 2\pi). \end{cases}$$

Using 2-homogeneity, we may compute the inverse map, on E_0, as

$$\sigma_J^{-1}(s) = \begin{cases} s + 4\pi, & s \in [-2\pi, -\frac{8\pi}{7}) \\ s + \frac{3\pi}{2}, & s \in [-\frac{8\pi}{7}, -\pi) \\ s - \frac{3\pi}{2}, & s \in [\pi, \frac{8\pi}{7}) \\ s - 4\pi, & s \in [\frac{8\pi}{7}, 2\pi). \end{cases}$$

Then σ_J^{-1} is not a 2π-congruence of E_0, so again, we have

$$U_{\sigma_J}^* \notin \mathcal{C}_{\widehat{\psi}_0}(\widehat{D}, \widehat{T}).$$

The compositions with $\sigma_{\frac{\pi}{7}}$ are not difficult to compute. On E_0, we have

$$(\sigma_{\frac{\pi}{7}} \circ \sigma_J)(s) = \begin{cases} s - 2\pi, & s \in [-2\pi, -\frac{10\pi}{7}) \\ s + 2\pi, & s \in [-\frac{10\pi}{7}, -\pi) \\ s, & s \in [\pi, \frac{8\pi}{7}) \\ s - 2\pi, & s \in [\frac{8\pi}{7}, \frac{10\pi}{7}) \\ s - 6\pi, & s \in [\frac{10\pi}{7}, 2\pi) \end{cases}$$

and

$$(\sigma_J \circ \sigma_{\frac{\pi}{7}})(s) = \begin{cases} s, & s \in [-2\pi, -\frac{12\pi}{7}) \\ s + 6\pi, & s \in [-\frac{12\pi}{7}, -\frac{10\pi}{7}) \\ s + 2\pi, & s \in [-\frac{10\pi}{7}, -\pi) \\ s + \pi, & s \in [\pi, \frac{8\pi}{7}) \\ s - 6\pi, & s \in [\frac{8\pi}{7}, 2\pi). \end{cases}$$

So $\sigma_{\frac{\pi}{7}} \circ \sigma_J$ *acts* 2π-congruently on E_0, but $\sigma_J \circ \sigma_{\frac{\pi}{7}}$ does not. Thus

$$U_{\sigma_{\frac{\pi}{7}}} U_{\sigma_J} \in \mathcal{C}_{\widehat{\psi}_0}(\widehat{D}, \widehat{T})$$

and

$$U_{\sigma_J} U_{\sigma_{\frac{\pi}{7}}} \notin \mathcal{C}_{\widehat{\psi}_0}(\widehat{D}, \widehat{T}).$$

Example A.8. Consider Example 4.5 (viii). Here, $A \subseteq [\pi, \frac{3\pi}{2})$ is a prescribed measurable subset, and B, C, D are derived from A so that $\{[\frac{3\pi}{2}, 2\pi), A, B, C, D\}$ partitions a wavelet set W. Let σ_W be the interpolation map between $E_0 = [-2\pi, -\pi) \cup [\pi, 2\pi)$ and W. The sets

$$\{[\frac{3\pi}{2}, 2\pi), \; A, \; B - 2\pi, \; C + 2\pi, \; D + 2\pi\}$$

partition $[0, 2\pi)$. So the sets

$$\{[\frac{3\pi}{2}, 2\pi), \; A, \; B - 4\pi, \; C + 2\pi, \; D\}$$

partition E_0. Hence

$$\sigma_W(s) = \begin{cases} s, & s \in [\frac{3\pi}{2}, 2\pi) \\ s, & s \in A \\ s + 4\pi, & s \in B - 4\pi \\ s - 2\pi, & s \in C + 2\pi \\ s, & s \in D \, . \end{cases}$$

From the construction we have $B = 2C + 4\pi$. So

$$\frac{1}{2}B = C + 2\pi \quad \text{and} \quad 2C = B - 4\pi.$$

So for $s \in B - 4\pi$, $\sigma_W(s) = s + 4\pi \in B$, hence

$$\sigma_W^2(s) = 2\sigma_W(\frac{1}{2}\sigma_W(s)) = 2(\frac{1}{2}\sigma_W(s) - 2\pi) = s.$$

Similarly, for $s \in C + 2\pi$, $\sigma_W(s) = s - 2\pi \in C$, so

$$\sigma_W^2(s) = \frac{1}{2}\sigma_W(2\sigma_W(s)) = \frac{1}{2}(2\sigma_W(s) + 4\pi) = s.$$

Thus the map σ_W is involutive.

Example A.9. We construct an example of an interpolation map σ between two wavelet sets E and F for which $\sigma^3 =$ identity, and σ^2 is a 2π-congruence on the initial wavelet set E. Hence

$$\text{Group}\{U_\sigma\} \subseteq \mathcal{C}_{\widehat{\psi}_E}(\widehat{D}, \widehat{T}),$$

so σ *admits* operator-interpolation.

Let

$$A = [-\pi, -\frac{\pi}{2}) \cup [\frac{7\pi}{2}, \frac{31\pi}{8}), \ B = [-\frac{\pi}{4}, -\frac{\pi}{8}) \cup [7\pi, \frac{31\pi}{2}).$$

Let

$$A_1 = [-\pi, -\frac{\pi}{2}), \ A_2 = [\frac{7\pi}{2}, \frac{15\pi}{4}), \ A_3 = [\frac{15\pi}{4}, \frac{31\pi}{8}),$$
$$B_1 = [-\frac{\pi}{4}, -\frac{\pi}{8}) \ B_2 = [7\pi, \frac{15\pi}{2}), \ B_3 = [\frac{15\pi}{2}, \frac{31\pi}{4}).$$

Then

$$A_1 + 8\pi = B_2, \ A_2 + 4\pi = B_3, \ A_3 - 4\pi = B_1,$$
$$A_1 = 4B_1, \ 2A_2 = B_2, \ 2A_3 = B_3.$$

It follows that A and B are both 2π-translation congruent and 2-dilation congruent. If fact, both are translation congruent to $[\pi, \frac{15\pi}{8})$ and both are dilation congruent to $[-2\pi, -\pi) \cup [\frac{7\pi}{4}, \frac{31\pi}{16})$. It follows, using an exhaustive induction technique, that there exists a (nonunique) measurable set C, disjoint from A and B, such that

$$E := A \cup C \text{ and } F := B \cup C$$

are wavelet sets. Let

$$\sigma := \sigma_E^F.$$

We have $\sigma(s) = s$ for $s \in C$, $\sigma(s) = s+8\pi \in B_2$ for $s \in A_1$, $\sigma(s) = s+4\pi \in B_3$ for $s \in A_2$, and $\sigma(s) = s - 4\pi \in B_1$ for $s \in A_3$. So on A_1, since $2A_2 = B_2$ we have $\frac{1}{2}\sigma(s) \in A_2$, hence

$$\sigma^2(s) = 2\sigma(\frac{1}{2}\sigma(s)) = 2(\frac{1}{2}\sigma(s) + 4\pi) = s + 16\pi.$$

Similar computations yield $\sigma^2(s) = s - 4\pi$ for $s \in A_2$, $\sigma^2(s) = s - 2\pi$ for $s \in A_3$, and of course $\sigma(s) = s$ for $s \in C$. So σ^2 is a 2π-congruence on E as required. To compute σ^3, note that for $s \in A_1$ we have $\sigma(s) \in B_2 = 2A_2$, so $\sigma^2(s) \in \sigma(2A_2) = 2\sigma(A_2) = 2B_3 = 4A_3$. Thus

$$\sigma^3(s) = 4\sigma(\frac{1}{4}\sigma^2(s)) = 4(\frac{1}{4}\sigma^2(s) - 4\pi) = \sigma^2(s) - 16\pi = s,$$

as required. For $s \in A_2$ we have $\sigma(s) \in B_3 = 2A_3$, so $\sigma^2(s) \in \sigma(2A_3) = 2B_1 = \frac{1}{2}A_1$, so

$$\sigma^3(s) = \frac{1}{2}\sigma(2\sigma^2(s)) = \frac{1}{2}(2\sigma^2(s) + 8\pi) = \sigma^2(s) + 4\pi = s.$$

For $s \in A_3$ we have $\sigma(s) \in B_1 = \frac{1}{4}A_1$, so $\sigma^2(s) \in \frac{1}{4}\sigma(A_1) = \frac{1}{4}B_2 = \frac{1}{2}A_2$, so

$$\sigma^3(s) = \frac{1}{2}\sigma(2\sigma^2(s)) = \frac{1}{2}(2\sigma^2(s) + 4\pi) = \sigma^2(s) + 2\pi = s.$$

Thus $\sigma^3 =$ identity on E, Hence on $\mathbb{R}$, as claimed.

Bibliography

[1] G. Battle, A block spin construction of ondelette, Part I: Lemarié functions, Comm. Math. Phys. 110 (1987), 601-615.

[2] J. Benedetto and M. Frazier (ed), *Wavelets: Mathematics and Applications,* CRC Press, Boca Raton, 1993.

[3] C. K. Chui, *An Introduction to Wavelets,* Acad. Press, New York, 1992.

[4] C. K. Chui (ed), *Wavelets: A Tutorial in Theory and Applications,* Acad. Press, New York, 1992.

[5] J. Conway, *A Course in Functional Analysis,* Springer-Verlag, New York, 1982.

[6] X. Dai and S. Lu, Wavelets in subspaces, the Michigan Math. J. Vol 43 (1996) 81-98.

[7] I. Daubechies, Orthonormal bases of compactly supported wavelets, Comm. Pure and Appl. Math. 41 (1988), 909-1005.

[8] I. Daubechies, *Ten Lectures on wavelets.,* CBMS 61, SIAM, 1992.

[9] I. Daubechies(ed.), *Different Perspectives on Wavelets,* Proc. of Symposia in Applied Math., vol. 47, Amer. Math. Soc., 1993.

[10] J. Dixmier, *von Neumann algebras,* North-Holland, 1981.

[11] T.N.T. Goodman, S.L. Lee and W.S. Tang, Wavelets in wandering subspaces, Trans. Amer. Math. Soc. 338 (1993) 639-654.

[12] T.N.T. Goodman, S.L. Lee and W.S. Tang, Wavelet bases for a set of commuting unitary operators, Adv. in Comp. Math. 1(1993), 109-126.

[13] P.R. Halmos, *A Hilbert Space Problem Book,* second ed., Springer-Verlag, New York, 1982.

[14] R. V. Kadison and J. R. Ringrose, *Fundamentals of the Theory of Operator Algebras.* vol. I and II, Acad. Press, New York, 1983, 1986.

[15] N. Kalouptsidis, M. Papadakis and T. Stavropoulos, An equivalence relation between multiresolution analyses of $L^2(\mathbb{R})$, preprint.

[16] Y. Katznelson, *An Introduction to Harmonic Analysis,* Dover, New York, 1976.

[17] D.R. Larson, Annihilators of operator algebras, Topics in Mod. Operator Theory 6, 119-130, Birkhauser Verlag, Basel, 1982.

[18] P. G. LeMarié , Ondelettes à localisation exponentielles , J. Math. Pure et Appl. 67 (1988), 227-236.

[19] S. Mallat, Multiresolution approximations and wavelet orthonormal basis of $L^2(\mathbb{R})$, Trans. Amer. Math. Soc. 315 (1989) 69-87.

[20] Y. Meyer, *Ondelettes et operateurs I,* Hermann editeurs des sciences et des arts, 1990; Eng. transl, *Wavelets and Operators,* Camb. Studies in Adv. Math. 37,1992.

[21] Y. Meyer, *Wavelets: Algorithms and Applications,* transl. from Fr., SIAM, Philadelphia, 1993.

[22] I. P. Natanson, *Theory of functions of a real variable,* transl. from Russ., F. Ungar, New York, 1961.

Editorial Information

To be published in the *Memoirs*, a paper must be correct, new, nontrivial, and significant. Further, it must be well written and of interest to a substantial number of mathematicians. Piecemeal results, such as an inconclusive step toward an unproved major theorem or a minor variation on a known result, are in general not acceptable for publication. *Transactions* Editors shall solicit and encourage publication of worthy papers. Papers appearing in *Memoirs* are generally longer than those appearing in *Transactions* with which it shares an editorial committee.

As of March 31, 1998, the backlog for this journal was approximately 9 volumes. This estimate is the result of dividing the number of manuscripts for this journal in the Providence office that have not yet gone to the printer on the above date by the average number of monographs per volume over the previous twelve months, reduced by the number of issues published in four months (the time necessary for preparing an issue for the printer). (There are 6 volumes per year, each containing at least 4 numbers.)

A Copyright Transfer Agreement is required before a paper will be published in this journal. By submitting a paper to this journal, authors certify that the manuscript has not been submitted to nor is it under consideration for publication by another journal, conference proceedings, or similar publication.

Information for Authors and Editors

Memoirs are printed by photo-offset from camera copy fully prepared by the author. This means that the finished book will look exactly like the copy submitted.

The paper must contain a *descriptive title* and an *abstract* that summarizes the article in language suitable for workers in the general field (algebra, analysis, etc.). The *descriptive title* should be short, but informative; useless or vague phrases such as "some remarks about" or "concerning" should be avoided. The *abstract* should be at least one complete sentence, and at most 300 words. Included with the footnotes to the paper, there should be the 1991 *Mathematics Subject Classification* representing the primary and secondary subjects of the article. This may be followed by a list of *key words and phrases* describing the subject matter of the article and taken from it. A list of the numbers may be found in the annual index of *Mathematical Reviews*, published with the December issue starting in 1990, as well as from the electronic service e-MATH [**telnet e-MATH.ams.org** (or **telnet 130.44.1.100**). Login and password are **e-math**]. For journal abbreviations used in bibliographies, see the list of serials in the latest *Mathematical Reviews* annual index. When the manuscript is submitted, authors should supply the editor with electronic addresses if available. These will be printed after the postal address at the end of each article.

Electronically prepared papers. The AMS encourages submission of electronically prepared papers in $\mathcal{A}_{\mathcal{M}}\mathcal{S}$-TEX or $\mathcal{A}_{\mathcal{M}}\mathcal{S}$-LATEX. The Society has prepared author packages for each AMS publication. Author packages include instructions for preparing electronic papers, the *AMS Author Handbook*, samples, and a style file that generates the particular design specifications of that publication series for both $\mathcal{A}_{\mathcal{M}}\mathcal{S}$-TEX and $\mathcal{A}_{\mathcal{M}}\mathcal{S}$-LATEX.

Authors with FTP access may retrieve an author package from the Society's Internet node **e-MATH.ams.org** (130.44.1.100). For those without FTP access, the author package can be obtained free of charge by sending e-mail to pub@ams.org (Internet) or from the Publication Division, American Mathematical Society, P.O. Box 6248, Providence, RI 02940-6248. When requesting an author package, please specify $\mathcal{AMS}$-TEX or $\mathcal{AMS}$-LATEX, Macintosh or IBM (3.5) format, and the publication in which your paper will appear. Please be sure to include your complete mailing address.

Submission of electronic files. At the time of submission, the source file(s) should be sent to the Providence office (this includes any TEX source file, any graphics files, and the DVI or PostScript file).

Before sending the source file, be sure you have proofread your paper carefully. The files you send must be the EXACT files used to generate the proof copy that was accepted for publication. For all publications, authors are required to send a printed copy of their paper, which exactly matches the copy approved for publication, along with any graphics that will appear in the paper.

TEX files may be submitted by email, FTP, or on diskette. The DVI file(s) and PostScript files should be submitted only by FTP or on diskette unless they are encoded properly to submit through e-mail. (DVI files are binary and PostScript files tend to be very large.)

Files sent by electronic mail should be addressed to the Internet address pub-submit@ams.org. The subject line of the message should include the publication code to identify it as a Memoir. TEX source files, DVI files, and PostScript files can be transferred over the Internet by FTP to the Internet node e-math.ams.org (130.44.1.100).

Electronic graphics. Figures may be submitted to the AMS in an electronic format. The AMS recommends that graphics created electronically be saved in Encapsulated PostScript (EPS) format. This includes graphics originated via a graphics application as well as scanned photographs or other computer-generated images.

If the graphics package used does not support EPS output, the graphics file should be saved in one of the standard graphics formats—such as TIFF, PICT, GIF, etc.—rather than in an application-dependent format. Graphics files submitted in an application-dependent format are not likely to be used. No matter what method was used to produce the graphic, it is necessary to provide a paper copy to the AMS.

Authors using graphics packages for the creation of electronic art should also avoid the use of any lines thinner than 0.5 points in width. Many graphics packages allow the user to specify a "hairline" for a very thin line. Hairlines often look acceptable when proofed on a typical laser printer. However, when produced on a high-resolution laser imagesetter, hairlines become nearly invisible and will be lost entirely in the final printing process.

Screens should be set to values between 15% and 85%. Screens which fall outside of this range are too light or too dark to print correctly.

Any inquiries concerning a paper that has been accepted for publication should be sent directly to the Editorial Department, American Mathematical Society, P. O. Box 6248, Providence, RI 02940-6248.

Selected Titles in This Series

(*Continued from the front of this publication*)

(See the AMS catalog for earlier titles)